Introduction to Time Series Modeling
with Applications in R
Second Edition

MONOGRAPHS ON STATISTICS AND APPLIED PROBABILITY

Editors: F. Bunea, R. Henderson, N. Keiding, L. Levina, R. Smith, W. Wong

Recently Published Titles

For more information about this series please visit: https://www.crcpress.com/
Chapman--HallCRC-Monographs-on-Statistics--Applied-Probability/
book-series/CHMONSTAAPP

Introduction to Time Series Modeling

with Applications in R

Second Edition

Genshiro Kitagawa

CRC Press
Taylor & Francis Group
Boca Raton London New York

CRC Press is an imprint of the
Taylor & Francis Group, an **informa** business

A CHAPMAN & HALL BOOK

Second edition published 2021
by CRC Press
6000 Broken Sound Parkway NW, Suite 300, Boca Raton, FL 33487-2742

and by CRC Press
2 Park Square, Milton Park, Abingdon, Oxon, OX14 4RN

jikeiretsu kaiseki nyumon
by Genshiro Kitagawa
© 2005 by Genshiro Kitagawa

Originally published in Japanese by Iwanami Shoten, Publishers, Tokyo, 2005.

This English language edition published in 2021
by Chapman & Hall/CRC, Boca Raton, FL, U.S.A.,
by arrangement with the author c/o Iwanami Shoten, Publishers, Tokyo

© 2021 Taylor & Francis Group, LLC
CRC Press is an imprint of Taylor & Francis Group, LLC

ISBN: 978-0-367-18733-0 (hbk)
ISBN: 978-0-429-19796-3 (ebk)

Contents

Preface

Due to rapid development of ICT (Information and Communication Technologies) and the progress of globalization, current societies have linked together in a complex manner at various levels, and there have arisen various difficult problems in economy, environment, health, safety, etc. Therefore, the elucidation, prediction and control of such dynamic complex systems are very important subjects. In statistical science, as a result of the development of the information criterion AIC, various methods of statistical modeling, in particular Bayes modeling, have developed. Further, with the advent of massive large-scale databases and fast parallel processors, many statistical models were developed and put into practice.

This book aims to introduce and explain basic methods of building models for time series. In time series modeling, we try to express the behavior of a certain phenomenon in relation to the past values of itself and other covariates. Since many important phenomena in statistical analysis are actually time series, and the identification of conditional distribution of phenomena is an essential part of the statistical modeling, it is very important and useful to learn basic methods of time series modeling. In this book, many time series models and various tools for handling them are introduced.

The main feature of this book is to use the state-space model as a generic tool for time series modeling. Three types of recursive filtering and smoothing methods—the Kalman filter, the non-Gaussian filter and the particle filter—are presented as convenient tools for the state-space models. Further, in this book, a unified approach to model evaluation is introduced based on the entropy maximization principle advocated by Dr. Akaike. Based on this unified approach, various methods of parameter estimation—such as the least squares method, the maximum likelihood method, the recursive estimation for the state-space models and the model selection by the information criterion AIC—are derived. After introducing standard stationary time series models such as AR model and ARMA model, we present various nonstationary time series models such as the locally stationary AR model, the trend model, the seasonal adjustment model, and the time-varying coefficient AR model and non-linear non-Gaussian models. The simulation methods are also shown. The principal aim of the author will be achieved when readers succeed in building their own models for their own real-world problems.

This book is basically the translation of the book published in Japanese in 2005 from Iwanami Publishing Company. The first version was published in 1993 as a volume in the Iwanami Computer Science Series. I would like to thank the Iwanami Publishing Company, in particular Mr. U. Yoshida, for allowing me to translate and publish in English.

I would like to acknowledge the many people who contributed to this book through collaborative research with the author. In particular, I would like to acknowledge with sincere thanks Hirotugu Akaike and Will Gersch, from whom I have learned so many ideas and the basis of time series modeling. Some of the models and estimation methods were developed during the process of cooperative research with Kohei Ohtsu, Tetsuo Takanami and Norio Matsumoto; they also provided me with some of the data used in this book. I have also been greatly influenced by discussions with S. Konishi, D. F. Findley, H. Tong, K. Tanabe, Y. Sakamoto, M. Ishiguro, Y. Ogata, Y. Tamura, T. Higuchi, Y. Kawasaki and S. Sato.

I am grateful to Prof. Subba Rao and two anonymous reviewers for their comments and suggestions that improved the original manuscript. I am also thankful to M. Oda for her help in preparing the manuscript of the English version. David Grubbs patiently encouraged and supported me throughout the preparation of this book. I express my sincere thanks to all of these people.

Genshiro Kitagawa
Tokyo, September 2009

Preface for Second Edition

The main difference of the second edition from the previous one is that it allows calculation of most of the modeling or methods explained in this book using the statistical language R. The R package TSSS (Time Series Analysis with State Space Model) was developed at the Institute of Statistical Mathematics, Japan, based on the FORTRAN programs that acompany the previous Japanese version of this book. Most of the examples in the book are computed by the functions in the package TSSS and the usage of the functions is shown as the examples. The package is available from CRAN and currently has 29 functions and ten sample data sets; most of them are real data. The revision of this book fully depends on the development of this package, and I sincerely thank Ms. Masami Saga and Prof. Junji Nakano for developing the TSSS package.

This revision was initiated based on the suggestion of David Grubbs, my editor at Chapman & Hall/CRC Press. In addition to adding R computations, we took this opportunity to make many additions and corrections throughout. Special thanks to him for giving me the opportunity to publish the revised edition. I also thank Ms. Hiromi Morikawa of Iwanami Publishing Company, Japan, for helping to comunicate with CRC Press.

Genshiro Kitagawa
Tokyo, January 2020

R and the Time Series Modeling Package TSSS

R is an open-source programming language for statistical computing and grphics which can run on Linux, Mac and Windows. Most of the data and analysis programs used as examples in this book are included in the TSSS (Time Series Analysis with State Space Model) package.

Getting R

To get R, go to the CRAN (Comprehensive R Archive Network) main page

https://cran.r-project.org/

where you can select the operating system: Linux, (Mac) OS X or Windows. There are nearly 100 mirror sites around the world.

The R Package TSSS

The TSSS package is also available from CRAN

https://cran.r-project.org/package=TSSS

TSSS was developed by Ms. Masami Saga and Prof. Junji Nakano at the Institute of Statistical Mathematics, Japan, based on the FORTRAN programs that accompany the previous Japanese version of this book.

The following functions are included in the package TSSS:

arfit	Univariate AR model fitting
armachar	Calculate characteristics of scalar ARMA model
armafit	Scalar ARMA model fitting
armafit2	Automatic scalar ARMA model fitting
boxcox	Box-Cox transformation
crscor	Cross-covariance and cross-correlation

fftper	Compute a periodogram via FFT
klinfo	Kullback-Leibler information
lsar	Decomposition of time interval to stationary subintervals
lsar.chgpt	Estimation of the change coint
lsqr	The least squares method via Householder transformation
marfit	Yule-Walker method of fitting multivariate AR model
marlsq	Least squares method for multivariate AR model
marspc	Cross-spectra and power contribution
ngsim	Simulation by non-Gaussian state-space model
ngsmth	Non-Gaussian smoothing
pdfunc	Probability density function
period	Compute a periodogram
pfilter	Particle filtering and smoothing
pfilterNL	Particle filtering and smoothing for nonlinear state-space model
polreg	Polynomial regression model
season	Seasonal adjustment
simssm	Simulation by Gaussian state-space model
trend	Trend estimation
tsmooth	Prediction and interpolation of time series
tvar	Time varying coefficients AR model
tvspc	Evolutionary power spectra by time varying AR model
tvvar	Time varying variance
unicor	Autocovariance and autocorrelation

The following data used as the example data are also included in the package TSSS:

BLSALLFOOD	Number of workers engaged in food industries
Haibara	Groundwater level data
HAKUSAN	Ship's navigation data
MYE1F	Seismic data
Nikkei225	Nikkei 225 stock price index data
NLmodel	Sample data for nonlinear state-space model
PfilterSample	Sample data for particle filter and smoother
Sunspot	Sunspot number data
Temperature	Temperatures data
WHARD	Wholesale hardware data

Chapter 1

Introduction and Preparatory Analysis

There are various types of time series, and it is very important to find out the characteristics of a time series by carefully looking at graphs of the data before proceeding to the modeling and analysis phase. In the first half of this chapter, preparatory analyses of time series such as drawing the graph of a time series, classification of time series from various viewpoints and the objectives of time series modeling considered in this book are discussed. In the second half of the chapter, we shall consider various ways of pre-processing time series that will be applied before proceeding to time series modeling. Finally, the organization of the book is described. Throughout this book, statistical computing and graphics environment R is used for obtaining the examples. The usage of R functions of the time series modeling package TSSS is also briefly introduced for each example.

1.1   Time Series Data

A record of phenomenon irregularly varying with time is called a *time series*. As examples of time series, we may consider meteorological data such as atmospheric pressure, temperature, rainfall and the records of seismic waves, economic data such as stock prices and exchange rates, medical data such as electroencephalograms and electrocardiograms, and records of controling cars, ships and aircraft.

As a first step in the analysis of a time series, it is important to carefully examine graphs of the data. These suggest necessary preprocessing before proceeding to the modeling and appropriate strategies for statistical modeling.

Throughout this book, we use the following data sets as examples. They are included in the TSSS package and can be taken into the R system by the function data as follows:

1

```
> data( HAKUSAN ) # Ship data
> data( Sunspot ) # Sunspot number data
> data( Temperature )# Tokyo daily maximum temperature data
> data( BLSALLFOOD ) # Number of emplyees in food industries
> data( WHARD ) # Monthly records of wholesale hardware data
> data( MYE1F ) # Seismic wave data
> data( Nikkei225 ) # Japanese stock price index data
> data( rainfall ) # Number of rainy days in 2 years data
> data( Haibara ) # Grandwater level data
```

To analyze other data sets not included in the package TSSS, it is possible to read in, for example, by the read_csv function. The function as.ts is used to create time series objects. In this function, we can specify the time of the first observation by the argument start and the number of observations per unit of time by frequency.

```
> sunspot <- as.ts( read.csv( "sunspot.csv" ) )
>
> blsfood <- as.ts( read.csv( "blsfood_new.csv" ), start =
c(1967,1), frequency = 12 )
```

Once the data set is taken into the R environment, the graph of the data is obtained by the function plot. In this function the arguments ylim and main are used to specify the range of the vertical axis and the title of the graph, respectively.

```
> hakusan <- as.ts( HAKUSAN[,1] )
> plot( hakusan,ylim=c(-8,8),main="(a) Ship's yaw rate data")
> plot( Sunspot,main="(b) Sunspot number data")
> plot( Temperature,main="(c) Maximum temperature data")
> plot( BLSALLFOOD,main="(d) BLSALLFOOD data")
> plot( WHARD,main="(e) WHARD data")
> plot( MYE1F,main="(f) Seismic data")
> plot( Nikkei225,main="(g) Stock price index data")
```

Figure 1.1 shows time series that will be analyzed in subsequent chapters as numerical examples. The following are the features of the time series shown in Figures 1.1 (a) – (i).

Plot (a) shows the first column of the HAKUSAN data, which is a time series of the *yaw rate* of a ship under navigation in the Pacific Ocean, observed every second. The yaw rate fluctuates around 0 degrees per second, because the ship is under the control of course keeping

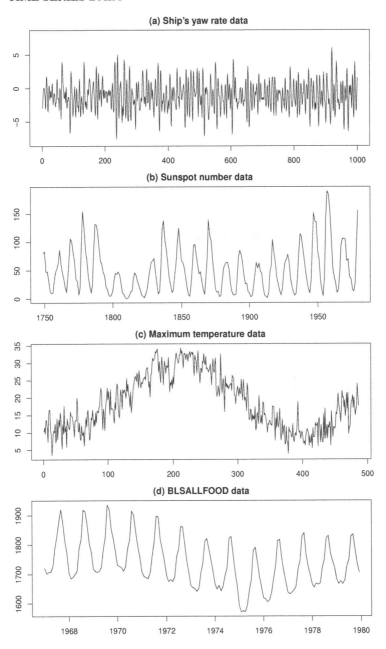

Figure 1.1: *Examples of various time series.*

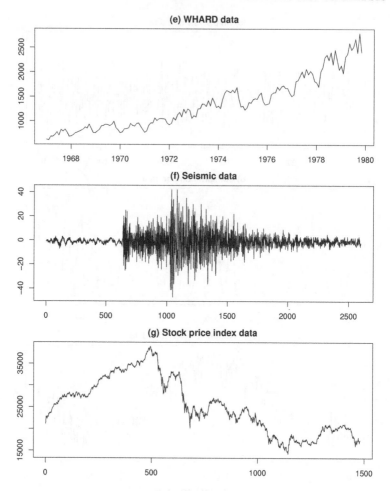

Figure 1.1: *Examples of various time series (continued).*

system (offered by Prof. K. Ohtsu of Tokyo University of Marine Science and Technology).

Plot (b) shows records of annual *sunspot numbers* (Wolfer sunspot number). Similar patterns of increase and decrease have been observed with an approximately eleven-year cycle.

Plot (c) shows the daily *maximum temperatures* for Tokyo recorded for 16 months. Irregular fluctuations around the predominant annual

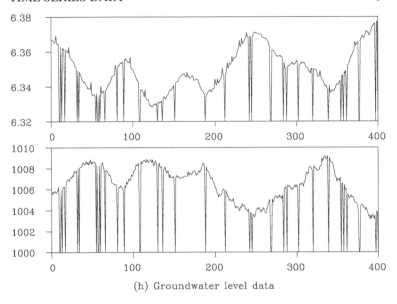

Figure 1.1: *Examples of various time series (continued).*

period (trend) are seen (source: Tokyo District Meteorological Observatory).

Plot (d) shows the monthly time series of the number of workers engaged in food industries in the United States, called the *BLSALL-FOOD data*. The data reveals typical features of economic time series that consist of trend and seasonal components. The trend component gradually varies, and the seasonal component repeats a similar annual pattern (source: the United States Bureau of Labor Statistics (BLS)).

Plot (e) shows the monthly record of wholesale hardware data, called *WHARD data*. This time series reveals typical characteristics of economic data that increase at an almost fixed rate every year, such that the fluctuations around the trend gradually increase in magnitude over time (source: US Bureau of Labor Statistics).

Plot (f) shows the time series of East-West components of *seismic waves*, recorded every 0.02 seconds. Due to the arrival of the P-wave (primary wave) and the S-wave (secondary wave), the variance of the series changed significantly. Moreover, it can be seen that not only the amplitude but also the frequency of the waves change with time (Takanami (1991)).

Plot (g) shows the daily closing values of the Japanese *stock price index*, Nikkei 225, quoted from January 4, 1988 to December 30, 1993. It reveals a monotone increase in values until the end of 1989, followed by a gradual decrease with large repetitive fluctuations after the bubble crash in Japan in the 1990s. In the analysis of stock price data, we often apply various analysis methods after taking the difference of the log-transformed data.

Multivariate time series

The R function plot can also draw multivariate time series.

```
> plot( Haibara )
>
> Hakusan <- as.ts(HAKUSAN[,c(2,3,4)])
> plot(Hakusan)
```

Plot (h) shows a bivariate time series of the *groundwater level* and the *atmospheric pressure* that were observed at ten-minute intervals at the observatory of the Tokai region, Japan, where a big earthquake was predicted to occur. In this plot, the observations that take the smallest value, i.e. 6.32, indicates the missing observations. On the other hand, the observations that markedly deviate upwards are considered as outliers that occurred due to malfunction of the observation device. To fully utilize the entire information contained in time series with many missing and outlying observations recorded over many years, it is necessary to develop an anlysis and modeling method that can be applied to data with such missing and outlying observations (offered by Dr. M. Takahashi and Dr. N. Matsumoto of National Institute of Advanced Industrial Science and Technology).

Plot (i) shows a multivariate time series of a ship's rolling, pitching and rudder angles recorded every second while navigating across the Pacific Ocean. As for the rolling and the rudder angles, both data show fluctuations over a cycle of approximately 16 seconds. On the other hand, the pitching angle varies over a shorter cycle of 10 seconds or less (offered by Prof. K. Ohtsu of Tokyo University of Marine Science and Technology).

1.2 Classification of Time Series

As has been shown in Figure 1.1, there is a wide variety of time series and they can be classified into several categories from various viewpoints.

(i) Hakusan

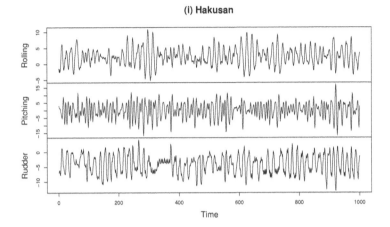

Figure 1.1: *Examples of various time series (continued).*

Continuous time series and discrete time series

Continuously recorded data, for example by an analog device, is called a *continuous* time series. On the other hand, data observed at a certain intervals of time, such as the atmospheric pressure measured hourly, is called a *discrete* time series.

There are two types of discrete time series. One is the *equally-spaced time series* and the other is the *unequally-spaced time series*. Although the time series shown in Figure 1.1 are connected continuously by solid lines, they are all discrete time series. Hereafter in this book, we consider only discrete time series recorded at equally spaced intervals, because time series that we analyze on digital computers are usually discrete time series.

Univariate and multivariate time series

Time series consisting of a single observation at each time point as shown in Figures 1.1(a) – 1.1(g) are called *univariate* time series. On the other hand, time series that are obtained by simultaneously recording two or more phenomena as the examples depicted in Figures 1.1(h) – 1.1(i) are called *multivariate* time series. However, it may be difficult to distinguish between univariate and multivariate time series from their nature; rather the distinction is made from the analyst's viewpoint and by various other factors, such as the measurement restriction and empirical or theoretical knowledge about the object. From a statistical modeling point of view, variable selection itself is an important problem in time series analysis.

Stationary and nonstationary time series

A time series is a record of a phenomenon irregularly varying over time. In time series analysis, irregularly varying time series are generally expressed by stochastic models. In some cases, a random phenomenon can be considered as a realization of a stochastic model with a time-invariant structure. Such a time series is called a *stationary* time series. Figure 1.1(a) is a typical example of a stationary time series.

On the other hand, if the stochastic structure of a time series itself changes over time, it is called a *nonstationary* time series. Typical examples of nonstationary time series are shown in Figures 1.1(c), (d), (e) and 1.1(g). It can be seen that mean values change over time in Figures 1.1(c), (d), (e) and 1.1(g), whereas the characteristics of the fluctuation around the mean value changes over time in Figure 1.1(f) and (g).

Gaussian and non-Gaussian time series

When a distribution of a time series follows a normal distribution, the time series is called a *Gaussian* time series; otherwise it is called a *non-Gaussian* time series. Most of the models considered in this book assume that the time series follow Gaussian distributions.

As in the case of Figure 1.1(b), the pattern of the time series is occasionally asymmetric so that the marginal distribution cannot be considered as Gaussian. Even in such a situation, we may consider that the time series is Gaussian after an appropriate data transformation. This method will be considered in Section 1.4 and Section 4.5.

Linear and nonlinear time series

A time series that is expressible as the output of a linear model, is called a *linear* time series. In contrast, the output from a nonlinear model is called a *nonlinear* time series.

Missing observations and outliers

In time series modeling of real-world problems, we sometimes need to deal with *missing observations* and *outliers*. A part of time series that have not been recorded for some reasons, are called missing observations; see Figure 1.1(h). Outliers (outlying observations) might occur due to extraordinary behavior of the object, malfunction of the observation device or errors in recording. In the groundwater level data shown in Figure 1.1(h), some data jumping upward are considered to be outliers.

1.3 Objectives of Time Series Analysis

This book presents statistical modeling methods for time series. The objectives of *time series analysis* considered in this book are classified into four categories; description, modeling, prediction and signal extraction.

Description: This includes methods that effectively express and visualize the characteristics of time series. By drawing figures of time series and basic descriptive statistics; such as sample autocorrelation functions, sample autocovariance functions and periodograms, we can capture essential characteristics of the time series, and get a hint for time series modeling.

Modeling: In time series modeling, we capture the stochastic structure of time series by identifying an appropriate model. Since there are various types of time series, it is necessary to select an adequate model class and to estimate parameters included in the model, depending on the characteristics of the time series and the objective of the time series analysis.

Prediction: In the prediction of time series, based on the correlations over time and among the variables, we can estimate the future behavior of time series by using various information extracted from current and past observations. In particular, in this book, we shall consider methods of prediction and simulation based on the estimated time series models.

Signal extraction: In signal extraction problems, we extract essential signals or useful information from time series corresponding to the objective of the analysis. To achieve that purpose, it is important to build models based on the salient characteristics of the time seires and the purpose of the analysis.

1.4 Pre-Processing of Time Series

For nonstationary time series, we sometimes perform pre-processing of the data before applying the analysis methods, which are introduced later in this book. This section treats some methods of transforming nonstationary time series into approximately stationary time series. In Chapter 11 and thereafter, however, we shall introduce various methods for modeling and analyzing nonstationary time series without pre-processing or stationalization.

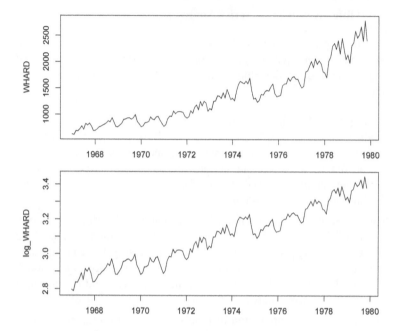

Figure 1.2: *WHARD data and log-transformed data.*

1.4.1 Transformation of variables

Some types of time series obtained by counting numbers or by measuring a positive-valued process such as the prices of goods and numbers of people illustrated in Figures 1.1(e) and 1.1(g), share the common characteristic that the variance of the series increases as the level of the series increases. For such a situation, by using the log-transformation $z_n = \log y_n$, we may obtain a new series whose variance is almost time-invariant and noise distribution is closer to the normal distribution than the original series y_n.

```
> data( WHARD )
> log_WHARD <- log10( WHARD )
> par( mfrow = c(2,1), mar = c(2,4,1,1)+0.1 )
> plot( WHARD, type = "l" )
> plot( log_WHARD, type = "l" )
```

Figure 1.2 shows the WHARD data and its log-transformed data. It can be seen that the log-transformed data is actually almost uniform and the trend of the series becomes almost linear.

A more general *Box-Cox transformation* (Box and Cox (1964)) includes the log-transformation as a special case and the automatic determination of its parameter will be considered later in Section 4.8. For time series y_n that take values in $(0, 1)$ like probabilities or ratios of the occurrence of a certain phenomenon, we can obtain a time series z_n that takes a value in $(-\infty, \infty)$ by the logit transformation

$$z_n = \log \left(\frac{y_n}{1 - y_n} \right). \tag{1.1}$$

In many cases, the distribution of the transformed time series z_n is less distorted than the original time series y_n, thus the modeling of the transformed series might be more tractable.

1.4.2   Differencing

When a time series y_n contains a trend as seen in Figures 1.1(c), (e) and (g), we often analyze the differenced series z_n defined by (Box and Jenkins (1970))

$$z_n = \Delta y_n = y_n - y_{n-1}. \tag{1.2}$$

This is motivated by the fact that if y_n is a straight line expressed as $y_n = a + bn$, then the differenced series z_n becomes a constant as

$$z_n = \Delta y_n = b, \tag{1.3}$$

and the slope of the straight line can be removed.

Furthermore, if y_n follows the quadratic function $y_n = a + bn + cn^2$, the difference of z_n becomes a constant and a and b are removed as follows:

$$
\begin{aligned}
\Delta z_n &= z_n - z_{n-1} \\
&= \Delta y_n - \Delta y_{n-1} \\
&= (b + 2cn) - (b + 2c(n-1)) \\
&= 2c. \tag{1.4}
\end{aligned}
$$

```
> plot(diff(log(nikkei225)))
```

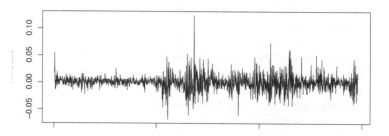

Figure 1.3: *Difference of the logarithm of the* Nikkei 225 *data.*

The R function `diff` returns the difference of the data. Figure 1.3 shows the difference of the logarithm, $r_n = \log y_n - \log y_{n-1}$, of the Nikkei 225 data shown in Figure 1.1(g), which is frequently used in the analysis of financial data in Japan. From Figure 1.3, it can be seen that the dispersion of the variation has changed abruptly around $n = 500$. We shall discuss this phenomenon in detail later in Sections 13.1 and 14.5.

When an annual cycle is observed in time series as shown in Figure 1.1(e), we would analyze the difference between the current value and the value at one cycle before, i.e.

$$\Delta_p y_n = y_n - y_{n-p}. \tag{1.5}$$

1.4.3   Month-to-month basis and year-over-year

For economic time series with prominent trend components as shown in Figure 1.1 (e), changes in the time series y_n from the previous month (or quarter) or year-over-year changes are often considered:

$$z_n = \frac{y_n}{y_{n-1}}, \qquad x_n = \frac{y_n}{y_{n-p}}. \tag{1.6}$$

where p is the cycle length or the number of observations in one year.

If the time series y_n is represented as the product of the trend T_n and the noise w_n as

$$y_n = T_n w_n, \tag{1.7}$$

and the trend component T_n grows by $T_n = (1 + \alpha)T_{n-1}$ where α is the growth rate, then the change from the previous month can be expressed by

$$z_n = \frac{y_n}{y_{n-1}} = \frac{T_n w_n}{T_{n-1} w_{n-1}} = (1 + \alpha)\frac{w_n}{w_{n-1}}. \tag{1.8}$$

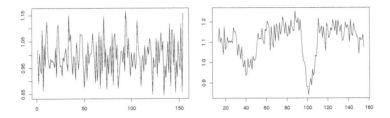

Figure 1.4 *Change from previous month (left) and year-over-year change (right) of WHARD data.*

This means that if the noise can be disregarded, the growth rate α can be determined by this transformation.

On the other hand, if y_n is represented as the product of a periodic function s_n with the cycle p and the noise w_n,

$$y_n = s_n \cdot w_n, \qquad s_n = s_{n-p}, \tag{1.9}$$

then the year-over-year x_n can be expressed by

$$x_n = \frac{y_n}{y_{n-p}} = \frac{s_n w_n}{s_{n-p} w_{n-p}} = \frac{w_n}{w_{n-p}}. \tag{1.10}$$

This suggests that the periodic function is removed by this transformation.

Example

The month-to-month basis (change from the previous month) and the year-over-year of the time series can be easily obtained by using the standard R function `lag` as follows:

```
> plot( whard/lag(whard) ) # change from previous month
> plot( whard/lag(whard,k=-12) ) # year-over-year change
```

Figure 1.4 shows the month-to-month basis and the year-over-year changes of the WHARD data shown in Figure 1.1(e). The trend component is removed in the month-to-month basis. On the other hand, the annual periodic component is removed in the year-over-year. By means of this year-over-year transformation, significant drops of the time series are revealed in the vicinity of $n = 40$ and $n = 100$.

1.4.4 Moving average

Moving average is a simple method for smoothing time series with random fluctuations. For a time series y_n, the $(2k+1)$-term moving average of y_n is defined by

$$T_n = \frac{1}{2k+1} \sum_{j=-k}^{k} y_{n+j}. \tag{1.11}$$

When the original time series is represented by the sum of the straight line t_n and the noise w_n as

$$y_n = t_n + w_n, \quad t_n = a + bn, \tag{1.12}$$

where w_n is an independent noise with mean 0 and variance σ^2, then the moving average becomes

$$T_n = t_n + \frac{1}{2k+1} \sum_{j=-k}^{k} w_{n+j}. \tag{1.13}$$

Here, since the sum of the independent noises satisfies

$$E\left[\sum_{j=-k}^{k} w_{n+j} \right] = \sum_{j=-k}^{k} E[w_{n+j}] = 0,$$

$$E\left[\left(\sum_{j=-k}^{k} w_{n+j} \right)^2 \right] = \sum_{j=-k}^{k} E\left[(w_{n+j})^2 \right] = (2k+1)\sigma^2, \tag{1.14}$$

the variance of the moving average becomes

$$\mathrm{Var}\left(\frac{1}{2k+1} \sum_{j=-k}^{k} w_{n+j} \right) = \frac{\sigma^2}{2k+1}. \tag{1.15}$$

This shows that the mean of the moving average T_n is the same as that of t_n and the variance is reduced to $1/(2k+1)$ of the original series y_n.

Example

Moving average of time series can be obtained the following code. Note that a similar result is obtained by the function SMA of the package TTR.

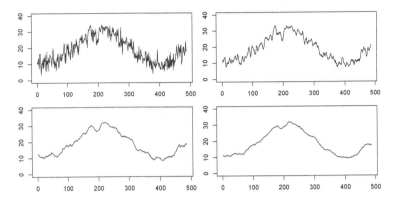

Figure 1.5 *Maximum temperature data and its moving average. Top left: original data, top right: moving average with $k = 2$, bottom left: $k = 8$, bottom right: $k = 14$.*

```
> data(Temperature)
> plot( Temperature,ylim=c(0,40) ) # original data
> y <- Temperature
> #
> # (2*kfilter+1)-terms moveing average filter
> #
> kfilter <- 2 # kfilter = 2,8,14
> ndata <- length( Temperature )
> y[1:ndata] <- NA
> for(i in 1:ndata){
> i0 <- max(i-kfilter,1)
> i1 <- min(i+kfilter,ndata)
> y[i] <- mean( Temperature[i0:i1] )
> }
> plot( y, ylim = c(0,40), type = "l")
```

Figure 1.5 shows the original maximum temperature data in Figure 1.1(c) and its moving averages with $k = 2, 8$ and 14. It can be seen that the moving averages yield smoother curves as k becomes larger.

In general, the weighted moving average is defined by

$$T_n = \sum_{j=-k}^{k} w_j y_{n-j}, \qquad (1.16)$$

where the weights satisfy $\sum_{j=-k}^{k} w_j = 1$ and $w_j \geq 0$.

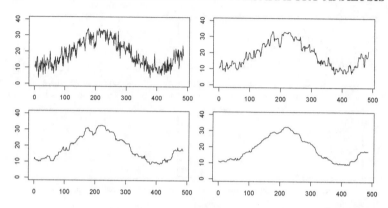

Figure 1.6 *Maximum temperature data and its moving median. Top left: original data, top right: moving median with $k = 2$, bottom left: $k = 8$, bottom right: $k = 14$.*

If we modify the definition of a moving average by using the median instead of the average, we obtain the $(2k+1)$-term moving median that is defined by

$$T_n = \text{median}\ \{y_{n-k}, \cdots, y_n, \cdots, y_{n+k}\}. \qquad (1.17)$$

The moving median can detect a change in the trend more quickly than the moving average.

To compute the moving median, it is only necessary to replace "mean" with "median" in the code for moving median, as follows:

```
> # (2*kfilter+1)-terms moveing median filter
> #
> for(i in 1:ndata){
> i0 <- max(i-kfilter,1)
> i1 <- min(i+kfilter,ndata)
> y[i] <- median( Temperature[i0:i1] )
> }
```

Figure 1.6 shows the original maximum temperature data in Figure 1.1(c) and its moving median for $k = 2, 8$ and 14. It can be seen that the moving median yield more robust estimates to the large deviation in the original data.

1.5   Organization of This Book

The main aim of this book is to provide basic tools for modeling various time series that arise for real-world problems. Chapters 2 and 3 are basic chapters and introduce two descriptive approaches. In Chapter 2, the autocovariance and autocorrelation functions are introduced as basic tools to describe univariate stationary time series. The cross-covariance and cross-correlation functions are also introduced for multivariate time series. In Chapter 3, the spectrum and the periodogram are introduced as basic tools for the frequency domain analysis of stationary time series. For multivariate case, the cross-spectrum and the power contribution are also introduced.

Chapters 4 and 5 discuss the basic methods for statistical modeling. In Chapter 4, typical probability distributions are introduced. Then, based on the entropy maximization principle, the likelihood function, the maximum likelihood method and the AIC criterion are derived. In Chapter 5, under the assumption of linearity and normality of the noise, the least squares method is derived as a convenient method for fitting various statistical models.

Chapters 6 to 8 are concerned with ARMA and AR models. In Chapter 6, the ARMA model is introduced, and the impulse response function, the autocovariance function, partial autocorrelation coefficients, the power spectrum and characteristic roots are derived from the ARMA model. The multivariate AR model is also considered in this chapter and the cross-spectrum and power contribution are derived. The Yule-Walker method and the least squares method for fitting an AR model are shown in Chapter 7. In Chapter 8, the AR model is extended to the case where the time series is piecewise stationary and an application of the model to the automatic determination of the change point of a time series is given.

Chapter 9 introduces the state-space model as a unified way of expressing stationary and nonstationary time series models. The Kalman filter and smoother are shown to provide the conditional mean and variance of the unknown state vector, given the observations. It is also shown that we can get a unified method for prediction, interpolation and parameter estimation by using the state-space model and the Kalman filter.

Chapters 10 to 13 show examples of the application of the state space model. In Chapter 10, the exact maximum likelihood method for the ARMA model is shown. The trend models are introduced in Chapter 11. In Chapter 12, the seasonal adjustment model is introduced to decompose seasonal time series into several components such as the trend and seasonal components. Chapter 13 is concerned with the modeling of nonstationarity in the variance and covariance. Time-varying coefficient AR models are introduced and applied to the estimation of a changing spectrum.

Chapters 14 and 15 are concerned with nonlinear non-Gaussian state-space models. In Chapter 14, the non-Gaussian state space model is introduced and a non-Gaussian filter and smoother are derived for state estimation. Applications to the detection of sudden changes of the trend component and other examples are presented. In Chapter 15, the particle filter and smoother are introduced as a very flexible method of filtering and smoothing for very general nonlinear non-Gaussian models.

Chapter 16 shows methods for generating various random numbers and time series that follow an arbitrarily specified time series model.

Algorithms for nonlinear optimaization and the particle filter and smoother and the derivations of the Levinson's algorithm and the Kalman filter are shown in Appendices.

Problems

1. What is necessary to consider when discretizing a continuous time series?

2. Give an example of a non-Gaussian time series and describe its characteristics.

3. (1) Obtain the inverse transformation of the logit transformation (1.1).

   (2) Find a transformation from (a,b) to $(-\infty,\infty)$ and find its inverse.

4. Describe the problem in constructing a stationary time series from a nonstationary time series by differencing.

5. Describe the problem in removing cyclic components by annual changes.

6. (1) Show that if the true trend is a straight line, then the mean value does not change for the three-term moving average, and that the variance becomes $1/3$ of the observed data variance.

   (2) Discuss the differences between the characteristics of the moving average filter and the moving median filter.

Chapter 2

The Covariance Function

In this chapter, the covariance and correlation functions are presented as basic methods to summarize and visualize stationary time series. The autocovariance function is a tool to express the relation between past and present values of time series and the cross-covariance function is to express the relation between two time series. These covariance functions are used to capture features of time series, to estimate the spectrum and to build time series models.

2.1   The Distribution of Time Series and Stationarity

The *mean* and the *variance* of data are frequently used as basic statistics to capture characteristics of random phenomena. A histogram is used to represent rough features of the data distribution. Therefore, by obtaining the mean, the variance and the histogram, it is expected to capture some important aspects or features of the data.

Accordingly, we shall investigate whether these statistics are useful for the analysis of time series. R code for drawing the histograms and the scatter plots between current and past values of a ship's data is as follows, where the argument par specifies the graphics parameters mfrow and mar

```
> data(HAKUSAN)
> par(mfrow=c(3,2))
> par(mar=c(2,2.5,3,1)+0.1)
> hist( HAKUSAN[,1], main="(a) Yaw rate" )
> hist( HAKUSAN[,2], main="(b) Rolling",breaks=seq(-6,12,1.2) )
> x <- as.ts(HAKUSAN[,2])
> y <- as.ts(HAKUSAN[,1])
> plot( x, lag(x,k=2), pch=20, main="(c) lag=2" )
> plot( y, lag(y,k=2), pch=20, main="(d) lag=2" )
> plot( x, lag(x,k=4), pch=20, main="(e) lag=4" )
> plot( y, lag(y,k=4), pch=20, main="(f) lag=4" )
```

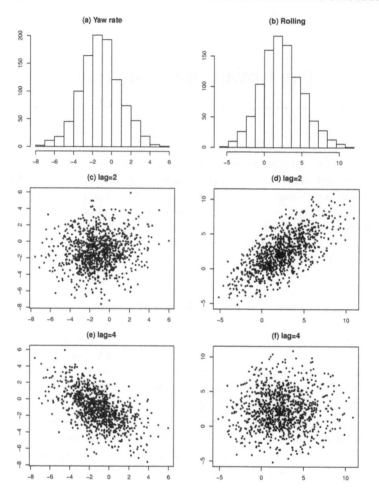

Figure 2.1: *Histograms and scatterplots of the yaw rate and rolling of a ship.*

The upper plots (a) and (b) in Figure 2.1 illustrate the histograms of the two time series; the ship's yaw rate and the rolling shown in Figure 1.1. However, although the time series in plot (a) and (i) of Figure 1.1 are apparently different, the histogram shown in Figure 2.1 (a) is quite similar to that of (b). This means that histograms cannot capture some aspects of the characteristics of the time series that are visually apparent.

Figures 2.1 (c) and (d) are *scatterplots* obtained by putting y_{n-2} on the horizontal axis and y_n on the vertical axis, for the ship's yaw rate and the roll data, respectively. Similarly, Figures 2.1(e) and (f) show scatterplots obtained by putting y_{n-4} on the horizontal axis, and y_n on the vertical axis.

The scatterplot in (c) shows that the data are distributed isotropically within a circle in the vicinity of the origin and this indicates that, in the case of the yaw rate, there is little correlation between y_n and y_{n-2}. On the other hand, the scatterplot shown in (d) is concentrated in a neighborhood of the diagonal, indicating that y_n has significant positive correlation with y_{n-2}. However, the scatterplot (e) concentrates on the negative diagonal, indicating a negative correlation between y_n and y_{n-4}. Such a correlation is not evident at all in the plot (f).

These examples show that in the analysis of time series, it is not possible to capture the essential features of time series by the marginal distribution of y_n that is obtained by ignoring the time series structure. Consequently, it is necessary to consider not only the distribution of y_n but also the joint distribution of y_n and y_{n-1}, y_n and y_{n-2}, and in general y_n and y_{n-k}. The properties of these joint distributions can be concisely expressed by the use of covariance and correlation coefficients of y_n and y_{n-k}.

Given a time series $y_1, \cdots, y_N$, the expectation of the time series y_n is defined by

$$\mu_n = E[y_n] \tag{2.1}$$

and is called the *mean value function*. Here $E[y]$ denotes the expectation of y. The covariance of a time series at two different times y_n and y_{n-k} is defined by

$$\mathrm{Cov}(y_n, y_{n-k}) = E[(y_n - \mu_n)(y_{n-k} - \mu_{n-k})], \tag{2.2}$$

and is called the *autocovariance* of the time series y_n (Box and Jenkins (1970), Brockwell and Davis (1991)). For $k = 0$, we obtain the variance of the time series at time n, $\mathrm{Var}(y_n)$.

In this chapter, we consider the case when the mean, the variance and the covariance do not change over time n. That is, we assume that for an arbitrary integer ℓ, it holds that

$$
\begin{array}{rcl}
E[y_n] & = & E[y_{n-\ell}] \\
\mathrm{Var}(y_n) & = & \mathrm{Var}(y_{n-\ell}) \\
\mathrm{Cov}(y_n, y_m) & = & \mathrm{Cov}(y_{n-\ell}, y_{m-\ell}).
\end{array}
\tag{2.3}
$$

A time series with these properties is called *weakly stationary* or *covariance stationary*. In Chapter 8 and later, sophisticated models are introduced for the analysis of general nonstationary time series for which the mean and the covariance change with time.

If the data are distributed as a normal (Gaussian) distribution, the characteristics of the distribution are completely determined by the mean, the variance and the covariance. However, such an assumption does not hold for many actual data. In such a situation, it is recommended to draw a histogram of the data. The histogram might reveal a difference in the distributions even though the two sets of data have the same mean, variance and covariance.

Therefore, the features of a time series cannot always be captured completely only by the mean, the variance and the covariance function. In general, it is necessary to examine the *joint probability density function* of the time series $y_1, \cdots, y_N$, i.e., $f(y_1, \cdots, y_N)$. For that purpose, it is sufficient to specify the joint probability density function $f(y_{i_1}, \cdots, y_{i_k})$ of $y_{i_1}, \cdots, y_{i_k}$ for arbitrary integers k and arbitrary time points satisfying $i_1 < i_2 < \cdots < i_k$.

In particular, when this joint distribution is a k-variate normal distribution, the time series is called a *Gaussian time series*. The features of a Gaussian time series can be completely captured by the mean vector and the variance covariance matrix.

When the distribution of a certain time series is invariant with respect to a time shift and the probability distribution does not change with time, the time series is called *strongly stationary*. Namely, a time series is called strongly stationary, if its probability density function satisfies the following relation

$$f(y_{i_1}, \cdots, y_{i_k}) = f(y_{i_1-\ell}, \cdots, y_{i_k-\ell}), \tag{2.4}$$

for any integer k, any time shift ℓ and any time points $i_1, \cdots, i_k$.

As noted above, the properties of Gaussian distributions are completely specified by the mean, the variance and the covariance. Therefore, for Gaussian time series, weak stationarity is equivalent to strong stationarity.

2.2 The Autocovariance Function of Stationary Time Series

Under the assumption of stationarity, the mean value function μ_n of a time series becomes a constant and does not depend on time n. Therefore, for a stationary time series, it can be expressed as

$$\mu = E[y_n], \tag{2.5}$$

where μ is called the *mean* of the time series y_n. Further, the covariance of y_n and y_{n-k}, $\text{Cov}(y_n, y_{n-k})$, becomes a value that depends only on the time difference k. Therefore, it can be expressed as

$$C_k = \text{Cov}(y_n, y_{n-k}) = \text{E}[(y_n - \mu)(y_{n-k} - \mu)], \qquad (2.6)$$

and is called the *autocovariance function* of the stationary time series (Box and Jenkins (1970), Brockwell and Davis (1991), Shumway and Stoffer (2000)). Here, k is called the *lag* or the time lag. When $k = 0$, the autocovariance is identical to the variance of y_n. The autocovariance function is an even function, i.e. $C_\ell = C_{-\ell}$, and it satisfies the inequality $|C_k| \le C_0$.

The correlation coefficient between y_n and y_{n-k} is defined by

$$R_k = \frac{\text{Cov}(y_n, y_{n-k})}{\sqrt{\text{Var}(y_n)\text{Var}(y_{n-k})}}, \qquad (2.7)$$

and by regarding it as a function of lag k, it is called the *autocorrelation function*.

For a stationary time series, since it holds that

$$\text{Var}(y_n) = \text{Var}(y_{n-k}) = C_0, \qquad (2.8)$$

the autocorrelation function is easily obtained from the autocovariance function as

$$R_k = \frac{C_k}{C_0}. \qquad (2.9)$$

Example (White noise)   When a time series y_n is a realization of an uncorrelated random variable with autocovariance function

$$C_k = \begin{cases} \sigma^2 & k = 0 \\ 0 & k \ne 0, \end{cases} \qquad (2.10)$$

it is called a *white noise* with variance σ^2. Obviously, the autocorrelation function of a white noise is given by $R_0 = 1$, $R_k = 0$ for $k = \pm 1, \pm 2 \dots$.

2.3   Estimation of the Autocovariance and Autocorrelation Functions

When a stationary time series $\{y_1, \cdots, y_N\}$ is given, the mean μ, the autocovariance function C_k and the autocorrelation function R_k are estimated by

$$\hat{\mu} \;=\; \frac{1}{N}\sum_{n=1}^{N} y_n \tag{2.11}$$

$$\hat{C}_k \;=\; \frac{1}{N}\sum_{n=k+1}^{N} (y_n - \hat{\mu})(y_{n-k} - \hat{\mu}) \tag{2.12}$$

$$\hat{R}_k \;=\; \frac{\hat{C}_k}{\hat{C}_0}, \tag{2.13}$$

respectively. Here $\hat{\mu}$ is called the *sample mean*, $\hat{C}_k$ is the *sample autocovariance function* and $\hat{R}_k$ is the *sample autocorrelation function*.

For Gaussian process, the variance of the sample autocorrelation $\hat{R}_k$ is given approximately by (Bartlett (1948), Box and Jenkins (1970))

$$\mathrm{var}(\hat{R}_k) \simeq \frac{1}{N}\sum_{j=-\infty}^{\infty} \left\{ R_j^2 + R_{j-k}R_{j+k} - 4R_k R_j R_{j-k} + 2R_j^2 R_k^2 \right\}. \tag{2.14}$$

Therefore, for time series for which all the autocorrelations R_j are zero for $j > m$ for some m, the variance of $\hat{R}_k$, $k > m$ is approximately given by

$$\mathrm{var}(\hat{R}_k) \simeq \frac{1}{N}\sum_{j=-\infty}^{\infty} R_j^2 \;=\; \frac{1}{N}\left(1 + 2\sum_{j=1}^{\infty} R_j^2 \right). \tag{2.15}$$

In particular, for a white noise sequence with $R_k = 0$ for $k > 0$, the approximate expression is simply given by

$$\mathrm{var}(\hat{R}_k) \simeq \frac{1}{N}. \tag{2.16}$$

This can be used for the test of whiteness of the time series. For example, the standard error of $\hat{R}_k$ is 0.1, 0.032 and 0.01 for $N = 100$, 1,000 and 10,000, respectively.

Example (Autocorrelation functions of time series)

Plots of autocorrelation functions R_k, $(k = 0,...,\text{lag})$ are obtained by the function unicor of the TSSS package as follows. In this function, if the lag is not explicitly specified, default is $2\sqrt{n}$ where n is the data length. Alternatively, the autcorrelation function can be also obtained by standard acf function in R.

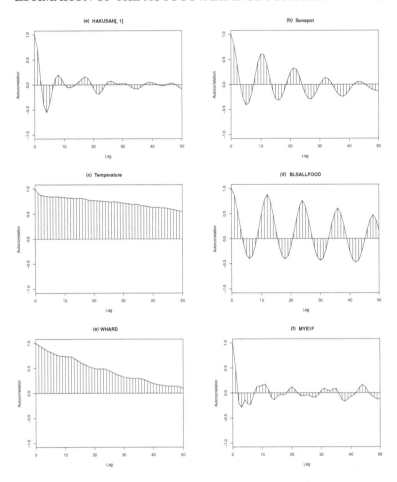

Figure 2.2: *Sample autocorrelation functions.*

```
> unicor(HAKUSAN[,1],lag=50)
> unicor(log(Sunspot),lag=50)
> unicor(Temperature,lag=50)
> unicor(BLSALLFOOD,lag=50)
> unicor(log(WHARD),lag=50)
> unicor(MYE1F,lag=50)
```

Figure 2.2 shows the sample autocorrelation functions of the time series shown in the Figure 1.1 (a) – (f). In the case of the stationary

time series of plot (a) of Figure 2.2, the peaks of the sample autocorrelation function rapidly decay to 0 with a cyclic fluctuation of period 8
or 9 as the lag increases. In the plot (b), the autocorrelation function of
the log-transformed series is illustrated, because the original data reveals
significant asymmetry. The peaks of the sample autocorrelation function repeatedly appear at an almost eleven-year cycle corresponding to
the approximate eleven-year cycle of the sunspot number data and the
amplitude of the sample autocorrelation gradually decreases as the lag
increases. The amplitude of the sample autocorrelation function in plot
(c) shows extremely slow decay, because a smooth annual trend is seen in
Figure 1.1 (c). These distinct features are common to most nonstationary
time series with a drifting mean value.

For the economic time series of plot (d), a one-year cycle in the sample autocorrelation function is seen corresponding to the annual cycle
of the time series. However, the amplitude of the sample autocorrelation
function decreases more slowly than those of (b) because of the presence of a trend in the time series. The economic time series of plot (e)
has an upward trend, and the amplitude grows along with it. Therefore,
the data have been log-transformed prior to computing the sample autocorrelation function. For the earthquake data of plot (f), the fluctuation
of the sample autocorrelation function continues for a considerably long
time with an approximate 10-second cycle after a sudden reduction in
the amplitude.

2.4   Multivariate Time Series and Scatterplots

Simultaneous records of random phenomena fluctuating over time are
called *multivariate time series*. An ℓ-variate time series is expressed as
$y_n = (y_n(1), \cdots, y_n(\ell))^T$, where $y_n(j)$, $j = 1, \cdots, \ell$, is the j-th time series
at time n, and y^T denotes the transpose of the vector y.

As discussed in the previous sections, the characteristics of a univariate time series are expressed by the autocovariance function and the
autocorrelation function. For multivariate time series, it is also necessary
to consider the relation between different variables.

As stated in the previous chapter, the first approach to time series
analysis is illustrating them with graphs. In the case of a multivariate
time series $y_n = (y_n(1), \cdots, y_n(\ell))^T$, the relation among the variables can
be understood by examining the *scatterplot*. The scatterplot of the time
series $y_n(i)$ versus $y_n(j)$ is obtained by plotting the point $(y_n(i), y_n(j))$ on
the two-dimensional plane with $y_n(i)$ shown on the horizontal axis and
$y_n(j)$ on the vertical axis.

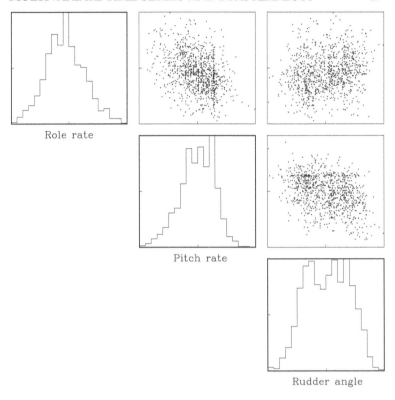

Role rate

Pitch rate

Rudder angle

Figure 2.3 *Histograms and scatterplots of three-dimensional ship's data, roll rate, pitch rate and rudder angle.*

Example

Standard R function `pairs` draws histogam and the scatterplot of multivariate data.

```
> pairs( HAKUSAN[,c(2,3,4)] )
```

In Figure 2.3, off-diagonal plots show the scatterplots of the three-variate ship's data (roll rate, pitch rate and rudder angle) that are shown in Figure 1.1 (i) and the diagonal plots show histograms of the time series $y_n(1), y_n(2)$ and $y_n(3)$.

Negative relations between two variables can be seen in the scatterplots of the roll rate and the pitch rate, and also of the pitch rate and the rudder angle. On the other hand, the scatterplot of the roll rate and the rudder angle are scattered over the whole region, which indicates that the simultaneous correlation is negligible between these two variables.

```
> pairs( Haibara )
```

Figure 2.4 shows the histograms and the scatterplot of the groundwater level data and the barometric pressure shown in Figure 1.1 (h). In the scatterplot, we see that the data points are concentrated near the negative diagonal. From the figures, we can see that the variation in the groundwater level corresponds closely to the variation in the barometric pressure.

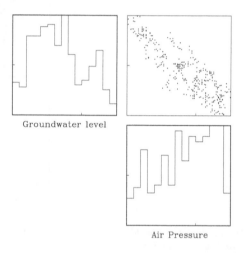

Figure 2.4 *Histograms and scatterplot of the groundwater level and barometric pressure data.*

As shown in the above examples, the relationships between two variables can be observed in the scatterplots. However, as was the situation with univariate time series, the relationship is not limited to the simultaneous case. Actually, for multivariate time series, we have to consider the relations between $y_n(i)$ and $y_{n-k}(j)$.

Therefore, to consider the relations among the multivariate time series, it is necessary to examine the scatterplots of $y_n(i)$ and $y_{n-k}(j)$ for all combinations of i, j, and k. To express such relations between variables with time delay very concisely, we introduce the cross-covariance and cross-correlation functions in the next section.

2.5   Cross-Covariance Function and Cross-Correlation Function

As stated in the previous sections, univariate time series are characterized by the three basic statistics, i.e. the mean, the autocovariance function and the autocorrelation function. Similarly, the mean vector, the cross-covariance function and the cross-correlation function are used to characterize the multivariate time series $y_n = (y_n(1), \cdots, y_n(\ell))^T$.

The mean of the i-th time series $y_n(i)$ is defined by,

$$\mu(i) = \mathrm{E}[y_n(i)], \qquad (2.17)$$

and $\mu = (\mu(1), \cdots, \mu(\ell))^T$ is called the *mean vector* of the multivariate time series y_n.

The covariance between the time series, $y_n(i)$, and the time series with time lag k, $y_{n-k}(j)$ is defined by

$$
\begin{aligned}
C_k(i,j) &= \mathrm{Cov}\big(y_n(i), y_{n-k}(j)\big) \\
&= \mathrm{E}\Big[\big(y_n(i) - \mu(i)\big)\big(y_{n-k}(j) - \mu(j)\big)^T\Big], \qquad (2.18)
\end{aligned}
$$

where the $\ell \times \ell$ matrix

$$
C_k = \begin{bmatrix}
C_k(1,1) & \cdots & C_k(1,\ell) \\
\vdots & \ddots & \vdots \\
C_k(\ell,1) & \cdots & C_k(\ell,\ell)
\end{bmatrix} \qquad (2.19)
$$

is called the cross-covariance matrix of lag k (Box and Jenkins (1970), Akaike and Nakagawa (1989), Brockwell and Davis (1991)).

Considering C_k, $k = 0, 1, 2, \cdots$, as a function of lag k, it is called a *cross-covariance function*. Here, the diagonal element $C_k(i,i)$ of the cross-covariance function is the autocovariance function of the i-th time series $y_n(i)$.

The correlation coefficient between $y_n(i)$ and $y_{n-k}(j)$ is defined by

$$
\begin{aligned}
R_k(i,j) &= \mathrm{Cor}(y_n(i), y_{n-k}(j)) \\
&= \frac{C_k(i,j)}{\sqrt{C_0(i,i)C_0(j,j)}}, \qquad (2.20)
\end{aligned}
$$

and the $\ell \times \ell$ matrix

$$
R_k = \begin{bmatrix}
R_k(1,1) & \cdots & R_k(1,\ell) \\
\vdots & \ddots & \vdots \\
R_k(\ell,1) & \cdots & R_k(\ell,\ell)
\end{bmatrix}, \qquad (2.21)
$$

is called the *cross-correlation function*.

The autocovariance function and the autocorrelation function are even functions and satisfy $C_{-k} = C_k$ and $R_{-k} = R_k$. For the multivariate time series, however, the cross-covariance function and the cross-correlation function do not have these symmetries in general. However, the relations

$$C_{-k} = C_k^T, \qquad R_{-k} = R_k^T \qquad (2.22)$$

do hold, so that it is sufficient to consider C_k and R_k only for $k \geq 0$.

When the multivariate time series $\{y_1(j), \cdots, y_N(j)\}$, $j = 1, \cdots, \ell$, of length N are given, estimates of the mean $\mu(i)$, the cross-covariance function $C_k(i, j)$ and the cross-correlation function $R_k(i, j)$ are obtained by

$$\hat{\mu}(i) \quad = \quad \frac{1}{N} \sum_{n=1}^{N} y_n(i) \qquad (2.23)$$

$$\hat{C}_k(i, j) \quad = \quad \frac{1}{N} \sum_{n=k+1}^{N} (y_n(i) - \hat{\mu}(i))(y_{n-k}(j) - \hat{\mu}(j)) \qquad (2.24)$$

$$\hat{R}_k(i, j) \quad = \quad \frac{\hat{C}_k(i, j)}{\sqrt{\hat{C}_0(i, i)\hat{C}_0(j, j)}}. \qquad (2.25)$$

Here, the ℓ-dimensional vector $\hat{\mu} = (\hat{\mu}(1), \ldots \hat{\mu}(\ell))^T$ is called the *sample mean vector*; the $\ell \times \ell$ matrix $\hat{C}_k = (\hat{C}_k(i, j))$ and $\hat{R}_k = (\hat{R}_k(i, j))$, $k = 0, 1, \ldots, i = 1, \ldots \ell, j = 1, \ldots \ell$, are called the *sample cross-covariance function* and the *sample cross-correlation function*, respectively.

Example (Ship data)   Cross-correlation function of a multivariate time series can be obtained by function crscor of the package TSSS as follows. In this function, if the lag is not explicitly specified, default is $2\sqrt{n}$ where n is the data length.

```
> data(HAKUSAN)
> y <- as.matrix(HAKUSAN[, 2:4]) # Rolling, Pitching, Rudder
> crscor(y, lag = 50)
```

Figure 2.5 shows the sample cross-correlation function of the ship's data consisting of roll rate, pitch rate and rudder angle as shown in Figure 1.1(i). The graph in the i-th row and the j-th column shows the correlation function $R_k(i, j)$, $k = 0, \cdots, 50$. The three plots on the diagonal of

Figure 2.5 show the autocorrelation function $R_k(i,i)$ and other six plots show the cross-correlation function $R_k(i,j)$.

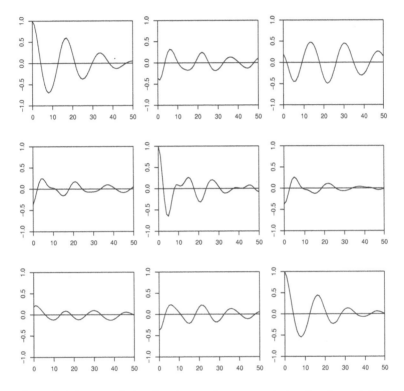

Figure 2.5 *Autocorrelation functions and cross-correlation functions of ship's data consisting of role rate, pitch rate and rudder angle.*

From these figures, it can be seen that the roll rate and the rudder angle fluctuate somewhat periodically. On the other hand, the autocorrelation function of the pitch rate is complicated. Furthermore, the plots in the figure indicate a strong correlation between the roll rate, $y_n(1)$, and the rudder angle, $y_{n-k}(3)$. Actually, the autocorrelation functions of the roll rate and rudder angle are similar and the cross-correlation $R_k(1,3)$ between the rudder angle and the roll rate is very high but $R_n(3,1)$ is rather small. The cross-correlation suggests that rudder angle affects the variation of the role rate but not vice versa.

Example (Groundwater level data)

```
> data(Haibara)
> crscor(Haibara,lag=50)
```

Figure 2.6 shows the cross-correlation function between the ground-water level data and the barometric pressure data illustrated in Figure 1.1(h). A very strong correlation between the two variables is seen.

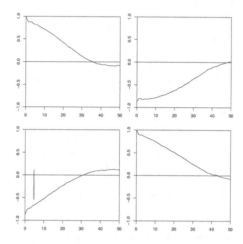

Figure 2.6 *Autocorrelation functions and cross-correlation functions of the groundwater level and barometric pressure data.*

Problems

1. Show that a weakly stationary Gaussian time series is strongly stationary.

2. Is a strongly stationary time series weakly stationary?

3. Show that the autocovariance function of a stationary time series is an even function.

4. Obtain the autocovariance function of a time series that satisfies $y_n = v_n - cv_{n-1}$ where $|c| < 1$ and v_n is a white noise with mean 0 and variance 1.

5. Assuming that $C_0, C_1, \ldots$, is the autocovariance function of a stationary time series, show that the matrix

$$
C = \begin{bmatrix}
C_0 & C_1 & \cdots & C_{k-1} \\
C_1 & C_0 & \ddots & \vdots \\
\vdots & \ddots & \ddots & C_1 \\
C_{k-1} & \cdots & C_1 & C_0
\end{bmatrix}
$$

is positive semi-definite.

6. (1) What is the expected value of the sample autocovariance function
$\hat{C}_k$?

(2) Discuss the reason that the sum is divided by N, not by $N - k$, in
the definition of the sample autocovariance function.

7. (1) Assuming that the time series is a Gaussian white noise, obtain
the distributions of $\hat{C}_k$ and $\hat{R}_k$.

(2) Using the results of (1), consider a method for checking whether
or not a time series is white.

Chapter 3

The Power Spectrum and the Periodogram

In this section, the spectral analysis method is introduced as a basic tool for stationary time series analysis. By means of spectral analysis, we can capture the characteristics of time series by decomposing time series into trigonometric functions at each frequency and by representing the features with the magnitude of each periodic component. The method discussed here will lead to the definition of the power spectrum and the periodogram of time series, computational methods, variance reduction and smoothing methods. Moreover, an efficient method of computing periodograms is presented using fast Fourier transforms (FFT). The readers interested in the spectral analysis of time series are referred to Brillinger (1974), Bloomfied (1976), Akaike and Nakagawa (1989) and Brockwell and Davis (1991).

3.1   The Power Spectrum

If the autocovariance function C_k rapidly decreases as the lag k increases and satisfies

$$\sum_{k=-\infty}^{\infty} |C_k| < \infty,$$

we can define the *Fourier transform* of C_k. The function defined on the frequency $-1/2 \leq f \leq 1/2$

$$p(f) = \sum_{k=-\infty}^{\infty} C_k e^{-2\pi i k f}, \tag{3.1}$$

is called the *power spectral density function* or simply the *power spectrum*.

Since the autocovariance function is an even function and satisfies $C_k = C_{-k}$, the power spectrum can also be expressed as

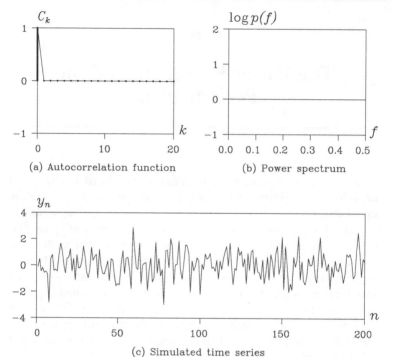

(a) Autocorrelation function

(b) Power spectrum

(c) Simulated time series

Figure 3.1 *Autocorrelation function, power spectrum and realization of a white noise with variance $\sigma^2 = 1$.*

$$
\begin{aligned}
p(f) &= \sum_{k=-\infty}^{\infty} C_k \cos(2\pi k f) - i \sum_{k=-\infty}^{\infty} C_k \sin(2\pi k f) \\
&= \sum_{k=-\infty}^{\infty} C_k \cos 2\pi k f = C_0 + 2 \sum_{k=1}^{\infty} C_k \cos 2\pi k f. \quad (3.2)
\end{aligned}
$$

The power spectrum represents a time series in term of trigonometric functions with various frequencies and expresses the characteristics of a time series by the magnitudes of these cyclic components. On the other hand, if a power spectrum is given, then the autocovariance function can be obtained via the inverse Fourier transform

$$
C_k = \int_{-\frac{1}{2}}^{\frac{1}{2}} p(f) e^{2\pi i k f} df = \int_{-\frac{1}{2}}^{\frac{1}{2}} p(f) \cos 2\pi k f \, df. \quad (3.3)
$$

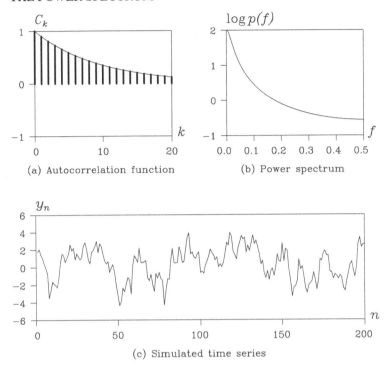

(a) Autocorrelation function   (b) Power spectrum

(c) Simulated time series

Figure 3.2 *Autocorrelation function, power spectrum and realization of a first-order AR model with a = 0.9.*

Example (The power spectrum of white noise)   The autocovariance function of a *white noise* is given by $C_0 = \sigma^2$ and $C_k = 0$ for $k \neq 0$. Therefore, the power spectrum of a white noise becomes

$$p(f) = \sum_{k=-\infty}^{\infty} C_k \cos 2\pi k f = C_0 = \sigma^2, \qquad (3.4)$$

taking a constant value for any frequency f. It means that a white noise contains cyclic components of various frequencies with the same magnitude. Plots (a), (b) and (c) in Figure 3.1 show the autocorrelation function and the power spectrum of a white noise with variance $\sigma^2 = 1$, and realizations of white noise generated by the simulation. The simulation method for a time series from an assumed model will be introduced later in Chapter 16.

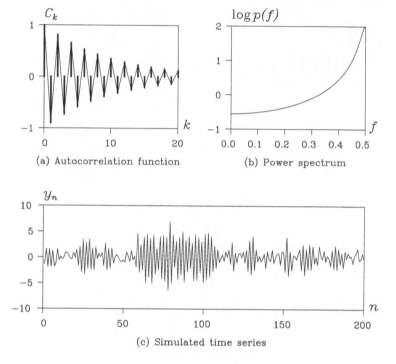

(a) Autocorrelation function   (b) Power spectrum

(c) Simulated time series

Figure 3.3 *Autocorrelation function, power spectrum and realization of a first-order AR model with* $a = -0.9$.

Example (Power spectrum of an AR model)   Assume that w_n is a white noise with variance σ^2. If a time series is generated by a first-order autoregressive (AR) model $y_n = ay_{n-1} + w_n$, the autocovariance function is given by $C_k = \sigma^2(1-a^2)^{-1}a^{|k|}$, and the power spectrum of this time series can be expressed as

$$p(f) = \frac{\sigma^2}{|1-ae^{-2\pi if}|^2} = \frac{\sigma^2}{1-2a\cos 2\pi f + a^2}. \qquad (3.5)$$

Note that autoregressive model will be explained in details in Chapter 6 and latter.

   Plots (a), (b) and (c) of Figure 3.2 show the autocorrelation function, the power spectrum and the realization, which were generated by the simulation for a first autoregressive model with $a = 0.9$. Similarly Figure 3.3 shows the case of a first-order autoregressive model with a negative coefficient $a = -0.9$. The autocorrelation function is very wig-

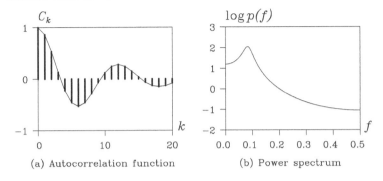

(a) Autocorrelation function   (b) Power spectrum

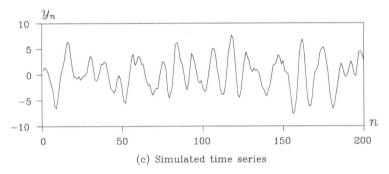

(c) Simulated time series

Figure 3.4 *Autocorrelation function, power spectrum and realization of a second-order AR model with $a_1 = 0.9\sqrt{3}$ and $a_2 = -0.81$.*

gly, and unlike the case of positive coefficient, the power spectrum is an increasing function of the frequency f.

If a time series follows a second-order AR model $y_n = a_1 y_{n-1} + a_2 y_{n-2} + w_n$, the autocorrelation function satisfies

$$R_1 = \frac{a_1}{1 - a_2}, \qquad R_k = a_1 R_{k-1} + a_2 R_{k-2}, \tag{3.6}$$

and the power spectrum can be expressed as

$$p(f) = \frac{\sigma^2}{|1 - a_1 e^{-2\pi i f} - a_2 e^{-4\pi i f}|^2}$$

$$= \frac{\sigma^2}{1 - 2a_1(1 - a_2)\cos 2\pi f - 2a_2 \cos 4\pi f + a_1^2 + a_2^2}. \tag{3.7}$$

Figure 3.4 shows (a) the autocorrelation function, (b) the power spectrum and (c) a realization, which were generated by the simulation with a second-order AR model with $a_1 = 0.9\sqrt{3}$ and $a_2 = -0.81$. The auto-correlation function is oscillatory, and the power spectrum has a peak around $f = 0.1$.

3.2   The Periodogram

Given a time series $y_1, \cdots, y_N$, the *periodogram* is defined by

$$p_j = \sum_{k=-N+1}^{N-1} \hat{C}_k e^{-2\pi i k f_j} = \hat{C}_0 + 2 \sum_{k=1}^{N-1} \hat{C}_k \cos 2\pi k f_j, \qquad (3.8)$$

where the sample autocovariance function $\hat{C}_k$ is substituted for the auto-covariance function C_k of equations (3.1) and (3.2).

In the definition of the periodogram, we consider only the natural frequencies defined by $f_j = j/N, j = 0, \cdots, [N/2]$. Here $[N/2]$ denotes the maximum integer, which does not exceed $N/2$. An extension of the periodogram

$$\hat{p}(f) = \sum_{k=-N+1}^{N-1} \hat{C}_k e^{-2\pi i k f}, \qquad -0.5 \le f \le 0.5, \qquad (3.9)$$

obtained by extending the domain to the continuous interval $[0, 1/2]$ is referred to as the *sample spectrum*.

In other words, the periodogram is obtained from the sample spectrum by restricting its domain to the natural frequencies. Corresponding to the relations of (3.3), the following relation holds between the sample spectrum and the sample autocovariance function,

$$\hat{C}_k = \int_{-\frac{1}{2}}^{\frac{1}{2}} \hat{p}(f) e^{2\pi i k f} df, \qquad k = 0, \cdots, N-1. \qquad (3.10)$$

Example (Periodograms)   In the R environment, the function `period` of the package TSSS yields the periodogram of a time series by setting `window=0`.

```
> period(HAKUSAN([,1]),window=0)
> period(Sunspot,window=0)
> period(Temperature,window=0)
> period(BLSALLFOOD,window=0)
> period(WHARD,window=0)
> period(MYE1F,window=0)
```

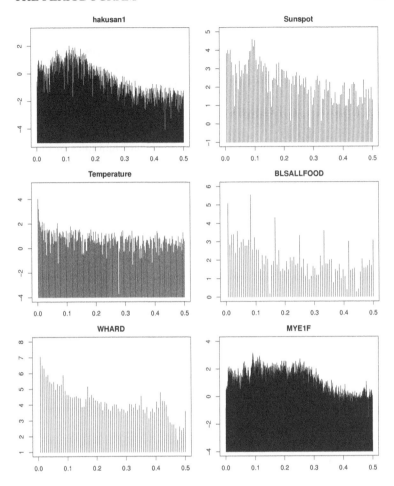

Figure 3.5: *Periodograms of the data of Figure 1.1 (on a logarithmic scale).*

Figure 3.5 shows the logarithm of the periodograms of the univariate time series introduced in Figures 1.1 (a) – (f).

Top-left plot shows the periodogram of the ship's (*Hakusan-maru*) yaw rate. Peaks of the periodograms are seen at about $f = 0.12$ and 0.07 that correspond to the cycle of 8.3 seconds and 14 seconds, respectively.

Top-right plot shows the periodogram of the annual sunspot number data. A strong periodic component with an approximate eleven-year cycle ($f = 0.09$) is shown. However, due to the strong variation, other periodic components cannot be clearly seen.

In the periodogram of the maximum temperature data shown in the middle-left plot, no apparent periodicity is seen except for the the zero frequency ($f = 0$), which corresponds to the trend of the data.

In the periodogram of the BLSALLFOOD data shown in the middle-right plot, seven sharp peaks are seen. In this periodogram the main peak at the frequency $f = 1/12$ corresponds to the seasonal components with twelve-month period, and other peaks are integer multiples of the frequency of the main peak, and are considered to be the higher harmonics of the nonlinear waveform.

Similar peaks can be seen in the WHARD data shown in the lower left plot, but they are not as prominent as the ones shown in the BLSALL-FOOD data.

The periodogram of the seismic data in the lower right plot has a plateau between $f = 0.1$ and $f = 0.25$, and two sharp peaks are seen at the frequencies $of = 0.07$ and $f = 0.1$.

Hereafter we shall briefly consider the properties of the periodogram. The periodogram and the sample spectrum are asymptotically unbiased and satisfy

$$\lim_{N\to\infty} \mathrm{E}\,[\hat{p}(f)] = p(f) = \sum_{k=-\infty}^{\infty} C_k \cos 2\pi k f. \qquad (3.11)$$

This means that at each frequency f, the expectation of the sample spectrum converges to the true spectrum as the number of data increases. However, it does not imply the consistency of $\hat{p}(f)$, that is, the sample spectrum $\hat{p}(f)$ does not necessarily converge to the true spectrum $p(f)$ as the number of data increases. Actually,

$$\frac{2\hat{p}(f_1)}{p(f_1)}, \ldots, \frac{2\hat{p}(f_{[\frac{N}{2}]-1})}{p(f_{[\frac{N}{2}]-1})}, \qquad (3.12)$$

follow the χ^2 distribution with two degrees of freedom independently, and $\hat{p}(0)/p(0)$ and $\hat{p}(0.5)/p(0.5)$ follow the χ^2 distribution with one degree of freedom. Therefore, the variance of the periodogram is a constant, independent of the sample size N. Thus the periodogram cannot be a consistent estimator.

Example (Sample autocorrelation functions and periodograms) In the R environment, the sample autocorrelation function and the periodogrm are obtained by using the functions unicor and period of the package TSSS, respectively. Three test data are generated by rnorm, which generates normal random numbers with sample size specified by the parameter lag.

```
> r <- as.ts(rnorm(3200))
> unicor(r,lag=50)
> x <- period(r,window=0)
> plot(log10(x$period),type="l",ylim=c(-4,1),xaxt="n")
> axis( side=1, at=c(0,320,640,960,1280,1600),
labels=c(0.0,0.1,0.2,0.3,0.4,0.5) )
>
> r1 <- r[1:800]
> unicor(r1,lag=50)
> x1 <- period(r1,window=0)
> plot(log10(x1$period),type="l",ylim=c(-4,1),xaxt="n")
axis( side=1, at=c(0,80,160,240,320,400),
labels=c(0.0,0.1,0.2,0.3,0.4,0.5) )
>
> r2 <- r[1:200]
> unicor(r2,lag=50)
> x2 <- period(r2,window=0)
> plot(log10(x2$period),type="l",ylim=c(-4,1),xaxt="n")
> axis( side=1, at=c(0,20,40,60,80,100),
labels=c(0.0,0.1,0.2,0.3,0.4,0.5) )
```

The upper plots in Figures 3.6 show the sample autocorrelation function and the periodogram, respectively, of the white noise with a sample size $N = 200$. Sample autocorrelations are close to zero and are almost contained in the confidence interval $[-1/\sqrt{200}, 1/\sqrt{200}] \simeq [-0.07, 0.07]$. On the other hand, the periodogram fluctuates sharply. Since the theoretical spectrum of the white noise is a constant, i.e., $\log p(f) \equiv 0$ in this case, this plot indicates that the periodogram is not a good estimate of the spectrum.

The middle and lower plots in Figures 3.6 show the sample autocorrelation functions and the periodograms of the white noise with sample sizes 800 and 3200, respectively. The sample autocorrelations $\hat{C}_k$ converge to the true value C_k as the data length increases. On the other hand, the amount of variation in the periodogram does not decrease even when the data length increases, and the frequency of the variation increases in proportion to N. This suggests that the sample spectrum will never converge to the true spectrum, no matter how many the sample size

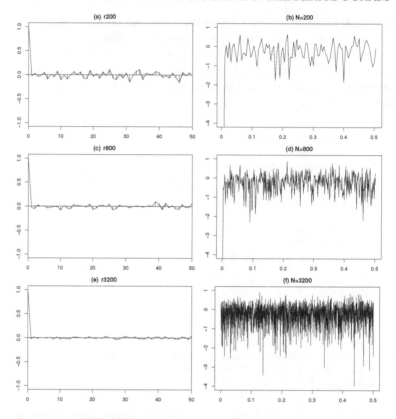

Figure 3.6 *Sample autocorrelation functions and periodograms in a logarithmic scale: sample sizes n = 200, 800, 3200.*

increases. This reflects the fact that the sample spectrum is not a consistent estimator of the spectrum.

3.3 Averaging and Smoothing of the Periodogram

In this section, we shall consider a method of obtaining an estimator that converges to the true spectrum as the data length n increases. Instead of (3.8), define p_j by

$$p_j = \hat{C}_0 + 2 \sum_{k=1}^{L-1} \hat{C}_k \cos 2\pi k f_j, \qquad (3.13)$$

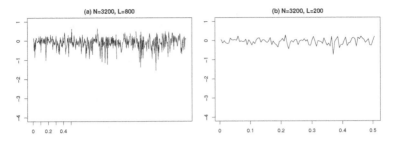

Figure 3.7 *Raw spectra of a white noise computed with the lag L = 200 and the sample size N = 800 (left plot) and N = 3,200 (right plot).*

for the frequencies $f_j = j/2L$ for $j = 0, \cdots, L$, where L is an arbitrary integer smaller than N.

The p_j defined in this way is called the *raw spectrum*. Figure 3.7 shows the raw spectrums p_j obtained by setting the lag to $L = 200$. The two plots are computed from the Gaussian white noise with $N = 800$ and 3200, respectively. Comparing with the top-right plot in Figure 3.6, which is obtained from $N = 200$ observations, it is seen that the fluctuation of the raw spectrum decreases as the data length N increases. This suggests that by using a fixed lag L instead of N, the raw spectrum p_j defined by (3.13) converges to the true spectrum.

```
> x <- period( r,window=0, lag=800 )
> plot(log10( x$period), type="l", ylim=c(-4,1), xaxt="n" )
> axis( side=1, at=c(0,80,160,240,320,400),
labels=c(0.0,0.1,0.2,0.3,0.4,0.5) )
>
> x <- period( r,window=0, lag=200 )
> plot( log10(x$period), type="l", ylim=c(-4,1), xaxt="n" )
> axis( side=1, at=c(0,80,160,240,320,400),
labels=c(0.0,0.1,0.2,0.3,0.4,0.5) )
```

In the definition of the periodogram (3.8), the sample autocovariances are necessary up to the lag $N - 1$. However, by just using the first $L - 1$ autocovariances for properly fixed lag L, the raw spectrum converges to the true spectrum as the data length (N) increases. On the assumption that the data length N increases according to $N = \ell L$, $\ell = 1, 2, \cdots$, computing the raw spectrum with the maximum lag $L - 1$ is equivalent to applying the following procedures.

Table 3.1 *Variances of the periodogram and the logarithm of the periodogram obtained from (3.8) and (3.14).*

Data length N	200	800	3200
p_j by (3.8)	1.006	0.998	1.010
p_j by (3.14)	1.006	0.250	0.061
$\log p_j$ by (3.8)	0.318	0.309	0.315
$\log p_j$ by (3.14)	0.318	0.053	0.012

Firstly, divide the time series $y_1, \cdots, y_n$ into N/L sub-series of length L, $y_j^{(i)}, \cdots, y_L^{(i)}$, $i = 1, \cdots, N/L$, namely, $y_j^{(i)} \equiv y_{(i-1)L+j}$, and a periodogram $p_j^{(i)}$, $j = 0, \cdots, [L/2]$ is obtained from each sub-series for $i = 1, \cdots, N/L$. After calculating the periodogram $p_j^{(i)}$, $j = 0, \cdots, [L/2]; i = 1, \cdots, N/L$, the *averaged periodogram* is obtained by averaging the N/L estimates for each $j = 0, \cdots, [L/2]$,

$$p_j = \frac{L}{N} \sum_{i=1}^{N/L} p_j^{(i)}. \tag{3.14}$$

By this procedure, the variance of each $p_j^{(i)}$ does not change, even if the number of data points, N, increases as $L, 2L, \cdots, \ell L$. However, since p_j is obtained as the mean of ℓ periodograms, $p_j^{(i)}$, $i = 1, \cdots, \ell$, the variance of p_j becomes $1/\ell$ of the variance of $p_j^{(i)}$. Therefore, the variance of p_j converges to 0 as the number of data points N or ℓ increases to infinity.

Table 3.1 shows the variances of the periodogram and the logarithm of the periodogram obtained by Eqs. (3.8) and (3.14), respectively. The variances of the periodogram obtained by Eq. (3.8) do not change as the number of data increases. Note that the theoretical variances of the periodogram and the log-periodogram are 1 and $\pi^2/6(\log 10)^2 \simeq 0.31025$, respectively. On the other hand, those obtained by (3.14) are inversely proportional to the data length. The reduction of the variances is also seen for the logarithm of the periodogram. In this case, the variances are reduced even faster.

Summarizing the above argument, although the periodogram does not converge to the true spectrum with the increase in the number of

Table 3.2: *Hanning and Hamming windows.*

Window	m	W_0	W_1
Hanning	1	0.50	0.25
Hamming	1	0.54	0.23

data, we can obtain an estimate of the spectrum that converges to the true spectrum by fixing the maximum lag, L, in computing the Fourier transform (3.13).

Here, note that the raw spectrum of (3.13) does not exactly agree with the averaged periodogram (3.14), and sometimes it might happen that $p_j < 0$. To prevent this situation and to guarantee the positivity of p_j, we need to compute p_j by (3.14). However, in actual computation, the raw spectrum is smoothed by using the *spectral window*. That is, for a given spectral window W_i, $i = 0, \pm 1, \cdots, \pm m$, an estimate of the spectrum is obtained by

$$\hat{p}_j = \sum_{i=-m}^{m} W_i p_{j-i}, \qquad j = 0, 1, \cdots, [L/2], \qquad (3.15)$$

where $p_{-j} = p_j$ and $p_{[L/2]+i} = p_{[L/2]-i}$. By properly selecting a spectral window, we can obtain an optimal estimate of the spectrum that is always positive and with smaller variance. Table 3.2 shows some typical spectral windows. Note that they are symmetric, $W_{-i} = W_i$, and also satisfy $W_i > 0$ to guarantee that $\hat{p}_j > 0$.

Example (Smoothed periodograms)

A smoothed periodogram is obtained by the function period. In the default setting, lag=$2\sqrt{N}$ and window =1. But we can also use other L and window type by using the parameters lag and window.

window: smoothing window type.
        0: box-car
        1: Hanning
        2: Hamming
lag: maximum lag of autocovariance. If Null (default).
        window=0: lag=$n-1$
        window>0: lag=$2\sqrt{n}$
        where n is the length of data

```
> period(HAKUSAN([,1]))
> period(Sunspot)
> period(Temperature)
> period(BLSALLFOOD)
> period(WHARD)
> period(MYE1F)
```

Figure 3.8 shows the smoothed periodograms obtained by putting $L = 2\sqrt{N}$ and applying the Hanning window to the data shown in Figure 3.5. When the number of data points is large, as occurs in Hakusan ship data and MYE1F data, the fluctuations become fairly small, so that reasonable estimates of the spectrum are obtained. However, in BLSALL-FOOD and WHARD data, the line spectra corresponding to the annual cycle become unclear due to the smoothing operation, although they can be clearly detected by the original periodogram as shown in Figure 3.5. Therefore, we need to make a compromise between the smoothness of the estimate and the sensitivity to the presence of significant peaks. This problem can be solved by efficient use of time series models as will be shown in Chapter 6.

3.4   Computational Method of Periodogram

The Görtzel method is known to be an effective method of calculating Fourier transforms exactly. If the Fourier cosine transform and the Fourier sine transform of the series $x_0, \cdots, x_{L-1}$ are computed directly,

$$X_c(f) = \sum_{n=0}^{L-1} x_n \cos(2\pi n f), \quad X_s(f) = \sum_{n=0}^{L-1} x_n \sin(2\pi n f),$$

L additions and multiplications and L evaluations of trigonometric functions are necessary for each calculation.

However, by adopting the Görtzel method, based on the additive theorem for trigonometric functions, to compute the Fourier transform $X_c(f)$ and $X_s(f)$, we only need to evaluate the trigonometric functions twice, i.e., $\cos 2\pi f$ and $\sin 2\pi f$. This algorithm is based on the following properties of the trigonometric functions. If we put $a_0 = 0$, $a_1 = 1$ and generate $a_2, \cdots, a_{L-1}$ by $a_n = 2a_{n-1} \cos 2\pi f - a_{n-2}$, then $\sin 2\pi n f$ and $\cos 2\pi n f$ can be obtained by

$$\sin 2\pi n f = a_n \sin 2\pi f, \quad \cos 2\pi n f = a_n \cos 2\pi f - a_{n-1},$$

respectively.

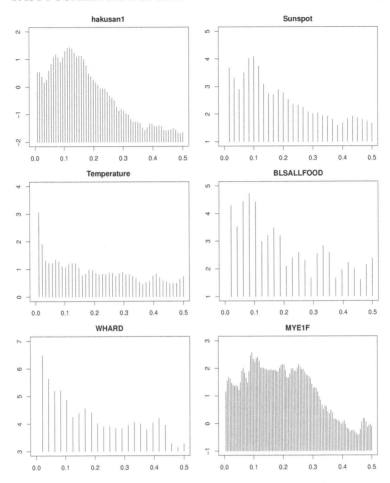

Figure 3.8 *Smoothed periodograms of the data shown in Figure 1.1. Horizontal axis: frequency f, vertical axis: periodogram on a logarithmic scale, $\log p(f)$.*

3.5   Computation of the Periodogram by Fast Fourier Transform

As noted in the last section, the periodogram is obtained from the discrete Fourier transform of the sample autocovariance function. In general, N^2 addition and multiplication operations are necessary to calculate the discrete Fourier transform of a time series of length N. Consequently, it takes a long time to compute the discrete Fourier transform when N is very large.

On the other hand, the fast Fourier transform (FFT) provides us with a very efficient algorithm. If the number of data points is of the form $N = p^\ell$, then this method requires approximately $Np\ell$ necessary operations, thus reducing the number of necessary operations by a factor of $Np\ell/N^2 = p\ell/N$. For instance, when $N = 2^{10} = 1024$ (or $N = 2^{20} \cong 10^6$), the number of necessary operations is reduced by a factor of approximately $1/50$ (or $1/25,000$).

However, if we calculate the periodogram by the formula (3.8), i.e. the Fourier transform of the sample autocovariance function, $N^2/2$ operations are required to obtain the sample autocovariance function $\hat{C}_k$, $k = 0, \cdots, N-1$. Therefore, it would be more efficient to apply the FFT algorithm directly to the time series to obtain the periodogram. The Fourier transform, X_j, of a time series $y_1, \cdots, y_N$ is obtained by

$$
\begin{aligned}
X_j &= \sum_{n=1}^{N} y_n e^{-2\pi i (n-1) j / N} \\
&= \sum_{n=1}^{N} y_n \cos \frac{2\pi(n-1)j}{N} - i \sum_{n=1}^{N} y_n \sin \frac{2\pi(n-1)j}{N} \\
&\equiv FC_j - i FS_j,
\end{aligned} \tag{3.16}
$$

for $j = 0, \cdots, N/L$.

Then the periodogram is obtained by

$$
p_j = \frac{|X_j|^2}{N} = \frac{1}{N} \left| \sum_{n=1}^{N} y_n e^{-2\pi i (n-1) j / N} \right|^2 = \frac{FC_j^2 + FS_j^2}{N}. \tag{3.17}
$$

It can be easily confirmed that the periodogram (3.17) agrees with the periodogram obtained using the original definition (3.8). In particular, when the length of the time series is $N = 2^\ell$ for some integer ℓ, the FFT is readily calculable.

For a time series of general length, i.e., not expressible in the form of $N = 2^\ell$, we might use the prime number factorization: $N = p_1^{\ell_1} \times \cdots \times p_m^{\ell_m}$. An alternative simple way of computing the periodogram by means of the FFT algorithm is to apply the FFT after modifying the time series by adding $(2^\ell - N)$ zeros behind the data, to make it of length $N' = 2^\ell$. By this method, we can obtain the same value for the sample spectrum that is calculated by equation (3.13) for the frequencies $f_j = j/N'$, $j = 0, \cdots, N'/2$.

It should be noted here that we can compute the Fourier transform for arbitrary frequencies f by using the original definition (3.13). However, if the data length is N, the frequencies of the periodogram obtained by the FFT algorithm are limited to $f_k = k/N$.

Therefore, if $N \neq 2^\ell$, the periodogram obtained using the FFT is evaluated at different frequencies from those of the periodogram obtained directly, using the definition in Section 3.2. These differences are not so crucial when the true spectrum is continuous. However, if the true spectrum contains line spectra or sharp peaks, the two methods might yield quite different results.

```
> r <- as.ts(rnorm(3200))
> # simulation of 2 cosine function + noise
> t <- 1:400
> r <- rnorm(400)
> y <- rep(0,400)
> for (i in t) {
> y[i] <- cos(2*pi*i/10) + cos(2*pi*i/4) + r[i]*0.1
> }
> y <- as.ts(y)
> plot(y,main="(a) Generated data")
```

Figure 3.9 (a) shows the realizations of the model that has two line spectra defined by

$$y_n = \cos \frac{2\pi n}{10} + \cos \frac{2\pi n}{4} + w_n, \qquad (3.18)$$

where $N = 400$ and $w_n \sim N(0, 0.01)$. Figure 3.9 (b) shows the periodogram obtained from the data $y_1, \cdots, y_{400}$ for which two line spectra are clearly detected at the frequencies $f = 0.05$ and $f = 0.0125$ corresponding to the two trigonometric functions of (3.18). For other frequencies which correspond to the white noise w_n, the periodogram fluctuates at a certain level.

```
> period(y,window=0)
> fftper(y,window=0)
```

On the other hand, Figure 3.9 (c) shows the periodogram obtained by using the FFT after generating the data with $N' = 512 = 2^9$ by adding 112 zeros behind $y_1, \cdots, y_{400}$. In this case, the periodogram (c) looks quite different from the periodogram (b), because the frequencies to be calculated for the periodogram (c) have deviated from the position of the line spectrum.

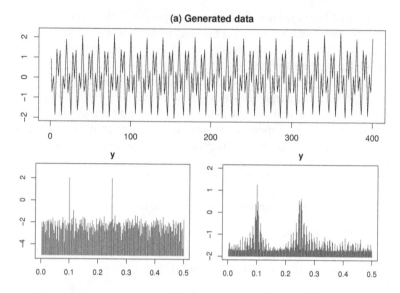

Figure 3.9 *Data with line spectrum and its periodograms obtained by (3.8) and by FFT after padding 112 zeros.*

```
> data(HAKUSAN)
> hakusan1 <- as.ts(HAKUSAN[,1])
> period(hakusan1,window=0)
> fftper(hakusan1,window=0)
```

Upper plot of Figure 3.10 duplicates the ship's yaw rate data shown in Figure 1.1 (a). The lower-left and lower-right plots show the periodograms obtained using the original definition and the FFT, respectively. This example shows that if the spectrum is continuous and does not contain any line spectra, the periodogram obtained by using FFT after padding zeros behind the data is similar to the periodogram obtained using the original definition.

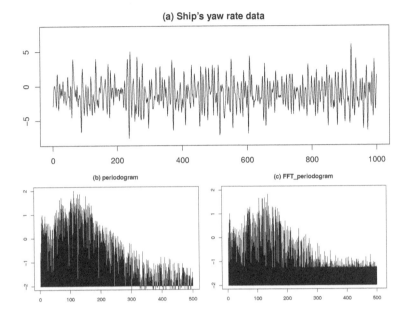

Figure 3.10: *Yaw rate data, periodogram and FFT spectrum.*

Problems

1. Verify equation (3.2), that the power spectrum can be expressed using a cosine function.

2. Using the results of problem 2.4 of Chapter 2, obtain the power spectrum of the time series $y_n = v_n - cv_{n-1}$, when v_n is a white noise with mean 0 and variance 1.

3. If the autocovariance function is given by $C_k = \sigma^2(1-a^2)^{-1}a^{|k|}$, verify that the power spectrum can be obtained using equation (3.5).

4. Show that the power spectrum of the time series (3.8) is given by (3.17).

5. Show the asymptotic unbiasedness of the sample spectrum, i.e. that $\lim_{n\to\infty} E\left[\hat{p}(f)\right] = p(f)$.

Chapter 4

Statistical Modeling

In the statistical analysis of time series, measurements of a phenomenon with uncertainty are considered to be the realization of a random variable that follows a certain probability distribution. Time series models and statistical models, in general, are built to specify this probability distribution based on data. In this chapter, a basic criterion is introduced for evaluating the closeness between the true probability distribution and the probability distribution specified by a model. Based on this criterion, we can derive a unified approach for building statistical models including the maximum likelihood method and the information criterion, AIC (Akaike (1973,1974), Sakamoto et al. (1986) and Konishi and Kitagawa (2008)).

4.1 Probability Distributions and Statistical Models

Given a *random variable* Y, assume that the probability of the event $Y \leq y$, $\mathrm{Prob}(Y \leq y)$, is defined for any real numbers $y \in R$. Considering this as a function of y, the function defined by

$$G(y) = \mathrm{Prob}(Y \leq y) \tag{4.1}$$

is called the *probability distribution function* (or *distribution function*) of the random variable Y.

Random variables used in time series analysis are usually *continuous*, and their distribution functions are expressible in integral form

$$G(y) = \int_{-\infty}^{y} g(t)dt, \tag{4.2}$$

with a function that satisfies $g(t) \geq 0$ for $-\infty < t < \infty$. Here, $g(x)$ is called a *density function*. On the other hand, if the distribution function or the density function is given, the probability that the random variable Y satisfies $a < Y \leq b$ for arbitrary $a < b$ is obtained by

$$G(b) - G(a) = \int_{a}^{b} g(x)dx. \tag{4.3}$$

In statistical analysis, various distributions are used to model characteristics of the data. Typical density functions are as follows.

(a) **Normal distribution (Gaussian distribution).** The distribution with density function

$$g(x) = \frac{1}{\sqrt{2\pi\sigma^2}} \exp\left\{-\frac{(x-\mu)^2}{2\sigma^2}\right\}, \qquad -\infty < x < \infty \qquad (4.4)$$

is called a *normal distribution* or a *Gaussian distribution* with the mean μ and the variance σ^2, and is denoted by $N(\mu, \sigma^2)$. In particular, $N(0, 1)$ is called the standard normal distribution.

(b) **Cauchy distribution.** The distribution with density function

$$g(x) = \frac{\tau}{\pi\{(x-\mu)^2 + \tau^2\}}, \qquad -\infty < x < \infty \qquad (4.5)$$

is called a *Cauchy distribution*. μ and τ^2 are called the location parameter and the dispersion parameter, respectively. Note that the square root of dispersion parameter, τ, is called the scale parameter.

(c) **Pearson family of distributions.** The distribution with density function

$$g(x) = \frac{c}{\{(x-\mu)^2 + \tau^2\}^b}, \qquad -\infty < x < \infty \qquad (4.6)$$

is called the (type VII) *Pearson family of distributions* with central parameter μ, dispersion parameter τ^2 and shape parameter b. Here c is a normalizing constant given by $c = \tau^{2b-1}\Gamma(b)/(\Gamma(b-\frac{1}{2})\Gamma(\frac{1}{2}))$. This distribution agrees with the Cauchy distribution for $b = 1$. Moreover, if $b = (k+1)/2$ with a positive integer k, it is called the *t-distribution* with k degrees of freedom.

(d) **Exponential distribution.** The distribution with density function

$$g(x) = \begin{cases} \lambda e^{-\lambda x} & \text{for } x \geq 0 \\ 0 & \text{for } x < 0, \end{cases} \qquad (4.7)$$

is called the exponential distribution. The mean and the variance of the exponential distribution are given by λ^{-1} and λ^{-2}, respectively.

(e) **χ^2 distribution (chi-square distribution).**
The distribution with density function

$$g(x) = \begin{cases} \dfrac{1}{2^{k/2}\Gamma(\frac{k}{2})} e^{-\frac{x}{2}} x^{\frac{k}{2}-1} & \text{for } x \geq 0 \\ 0 & \text{for } x < 0, \end{cases} \qquad (4.8)$$

is called the χ^2 *distribution* with k degrees of freedom. Especially, for $k = 2$, it becomes an exponential distribution. The mean and the variance of the χ^2 distribution are k and $2k$, respectively. The sum of the square of k Gaussian random variables follows the χ^2 distribution with k degrees of freedom.

(f) **Double exponential distribution.** The distribution with density function

$$g(x) = \exp\{x - e^x\} \tag{4.9}$$

is called the *double exponential distribution*. The mean and the variance of this distribution are $-\zeta$ and $\pi^2/6$, respectively, where $\zeta = 0.577224$ is the Euler constant. The logarithm of the exponential random variable follows the double exponential distribution.

(g) **Uniform distribution.** The distribution with density function

$$g(x) = \begin{cases} (b-a)^{-1}, & \text{for } a \le x < b \\ 0, & \text{otherwise} \end{cases} \tag{4.10}$$

is called the *uniform distribution* over $[a,b)$. The mean and the variance of the uniform distribution are $(a+b)/2$ and $(b-a)^2/12$, respectively.

Example The above-mentioned density functions can be drawn by the function pdfunc of the TSSS package, where xmin and xmax indicate the lower and upper bound of the domain of the density function, respectively.

```
> par(mar=c(2,2,3,1)+0.1)
> # normal distribution
> pdfunc(model = "normal", xmin = -4, xmax = 4)
> # Cauchy distribution
> pdfunc(model = "Cauchy", xmin = -4, xmax = 4)
> # Pearson distribution
> pdfunc(model = "Pearson", shape = 2, xmin = -4, xmax = 4)
> # exponential distribution
> pdfunc(model = "exp", xmin = 0, xmax = 8)
> # Chi-square distribution
> pdfunc(model = "Chi2", df = 3, xmin = 0, xmax = 8)
> # double exponential distribution
> pdfunc(model = "dexp", xmin = -4, xmax = 4)
```

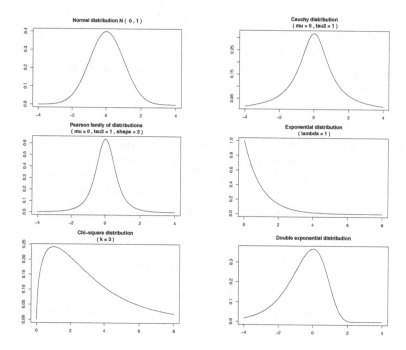

Figure 4.1 *Density functions of various probability distributions. Normal distri-bution, Cauchy distribution, Pearson distribution, exponential distribution, χ^2 distribution and double exponential distribution are shown in order from upper-left panel to the lower-right panel.*

Figure 4.1 shows the density functions defined in (a)–(f) above. By the simulation methods to be discussed in Chapter 16, we can generate data $y_1, \cdots, y_N$ that follow these density functions. The generated data are referred to as *realizations* of the random variable. Figure 4.2 shows examples of realizations with the sample size $N = 20$ for the distributions of (a)–(c) and (f) above.

As mentioned above, if a probability distribution or a density function is given, we can generate data that follow the distribution. On the other hand, in statistical analysis, when data $y_1, \cdots, y_N$ are given, they are considered to be realizations of a random variable Y. That is, we assume a random variable Y underlying the data and consider the data

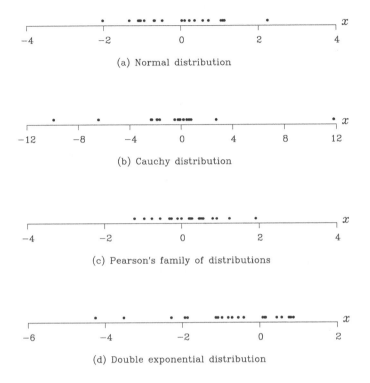

Figure 4.2: *Realizations of various probability distributions.*

as realizations of that random variable. Here, the density function $g(y)$ defining the random variable is called the *true model*. Since this true model is usually unknown for us, it is necessary to estimate the probability distribution from the data. For example, we estimate the density function shown in Figure 4.1 from the data shown in Figure 4.2. Here, the density function estimated from data is called a *statistical model* and is denoted by $f(y)$.

In ordinary statistical analysis, the probability distribution is sufficient to characterize the data, whereas for time series data, we have to consider the joint distribution $f(y_1, \cdots, y_N)$ as shown in Chapter 2. In Chapter 2, we characterized the time series $y_1, \cdots, y_N$ using the sample mean $\hat{\mu}$ and the sample autocovariance function $\hat{C}_k$. The implicit assumption behind this is that the N-dimensional vector $y = (y_1, \cdots, y_N)^T$

follows a multidimensional normal distribution with mean vector $\hat{\mu} = (\hat{\mu}, \cdots, \hat{\mu})^T$ and variance covariance matrix

$$
\hat{C} = \begin{bmatrix} \hat{C}_0 & \hat{C}_1 & \cdots & \hat{C}_{N-1} \\ \hat{C}_1 & \hat{C}_0 & \cdots & \hat{C}_{N-2} \\ \vdots & \vdots & \ddots & \vdots \\ \hat{C}_{N-1} & \hat{C}_{N-2} & \cdots & \hat{C}_0 \end{bmatrix}.
\tag{4.11}
$$

This model can express an arbitrary Gaussian stationary time series very flexibly. However, it cannot achieve an efficient expression of the information contained in the data since it requires the estimation of $N + 1$ unknown parameters, $\hat{C}_0, \cdots, \hat{C}_{N-1}$ and $\hat{\mu}$, from N observations. On the other hand, stationary time series models that will be discussed in Chapter 5 and later can express the covariance matrix of (4.11) using only a small number of parameters.

4.2   K-L Information and Entropy Maximization Principle

It is assumed that a true model generating the data is $g(y)$ and that $f(y)$ is an approximating statistical model. In statistical modeling, we aim at building a model $f(y)$ that is "close" to the true model $g(x)$. To achieve this, it is necessary to define a criterion to evaluate the goodness of the model $f(y)$ objectively.

In this book, we use the *Kullback-Leibler information* (hereinafter, abbreviated as *K-L information* (Kullback and Leibler (1951)))

$$
I(g;f) = E_Y \left[\log \left\{ \frac{g(Y)}{f(Y)} \right\} \right] = \int_{-\infty}^{\infty} \log \left\{ \frac{g(y)}{f(y)} \right\} g(y) dy
\tag{4.12}
$$

as a criterion. Here, E_Y denotes the expectation with respect to the true density function $g(y)$ and the last expression in (4.12) applies to a model with a continuous probability distribution. This K-L information has the following properties:

(i) $I(g;f) \geq 0$

(ii) $I(g;f) = 0 \iff g(y) = f(y).$ (4.13)

The negative of the K-L information, $B(g;f) = -I(g;f)$, is called the generalized (or Boltzmann) entropy. When n realizations are obtained from the model distribution $f(y)$, the entropy is approximately $1/N$ of the logarithm of the probability that the relative frequency distribution coincides with the true distribution $g(y)$. Therefore, we can say that the smaller the value of the K-L information, the closer the probability distribution $f(y)$ is to the true distribution $g(y)$. Statistical models approximate the true distribution $g(y)$ based on the data $y_1, \cdots, y_N$, whose goodness of approximation can be evaluated by the K-L information, $I(g;f)$. In statistical modeling, the strategy of constructing a model so as to maximize the entropy $B(g;f) = -I(g;f)$ is referred to as the *entropy maximization principle* (Akaike (1977)).

Example (K-L information of a normal distribution model)   Consider the case where both the true model, $g(y)$, and the approximate model, $f(y)$, are normal distributions defined by

$$
\begin{aligned}
g(y|\mu, \sigma^2) &= \frac{1}{\sqrt{2\pi\sigma^2}} \exp\left\{-\frac{(y-\mu)^2}{2\sigma^2}\right\} \\
f(y|\xi, \tau^2) &= \frac{1}{\sqrt{2\pi\tau^2}} \exp\left\{-\frac{(y-\xi)^2}{2\tau^2}\right\}.
\end{aligned} \tag{4.14}
$$

In this case, since the following expression holds:

$$
\log\left\{\frac{g(y)}{f(y)}\right\} = \frac{1}{2}\left\{\log\frac{\tau^2}{\sigma^2} - \frac{(y-\mu)^2}{\sigma^2} + \frac{(y-\xi)^2}{\tau^2}\right\}, \tag{4.15}
$$

the K-L information is given by

$$
\begin{aligned}
I(g;f) &= E_Y \log\left\{\frac{g(Y)}{f(Y)}\right\} \\
&= \frac{1}{2}\left\{\log\frac{\tau^2}{\sigma^2} - \frac{E_Y(Y-\mu)^2}{\sigma^2} + \frac{E_Y(Y-\xi)^2}{\tau^2}\right\} \\
&= \frac{1}{2}\left\{\log\frac{\tau^2}{\sigma^2} - 1 + \frac{\sigma^2 + (\mu-\xi)^2}{\tau^2}\right\}. \tag{4.16}
\end{aligned}
$$

For example, if the true distribution $g(y)$ is the standard normal distribution, $N(0,1)$, and the model $f(x)$ is $N(0.1, 1.5)$, then the K-L information can be easily evaluated as $I(g;f) = (\log 1.5 - 1 + 1.01/1.5)/2 = 0.03940$.

As shown in the example above, the K-L information $I(g;f)$ is easily calculated, if both g and f are normal distributions. However, for general distributions g and f, it is not always possible to compute $I(g;f)$ analytically. Therefore, in general, to obtain the K-L information, we need to resort to numerical computation. To illustrate the accuracy of numerical computation, Table 4.1 shows the K-L information with respect to two density functions, $g(y)$ and $f(y)$, obtained by numerical integration over $[x_0, x_k]$ using the trapezoidal rule

$$\hat{I}(g;f) = \frac{\Delta x}{2} \sum_{i=1}^{k} \{h(x_i) + h(x_{i-1})\}, \qquad (4.17)$$

where k is the number of nodes and

$$
\begin{aligned}
x_0 &= -x_k, \\
x_i &= x_0 + (x_k - x_0)\frac{i}{k} \qquad (4.18) \\
h(x) &= g(x)\log\frac{g(x)}{f(x)} \qquad (4.19) \\
\Delta x &= \frac{x_k - x_0}{k}.
\end{aligned}
$$

The function `klinfo` of the package TSSS computes the K-L information of the distribution $f(x)$ (normal distribution if `distf=1` and Cauchy distribution if `distf=2`) with respect to the true distribution $g(x)$ (normal distribution if `distg=1` and Cauchy distribution if `distg=2`). The parameters of the distributions are specified by `paramf` and `paramg`, respectively. The upper limit of the domain of integration is specified by `xmax`, and the lower limit is automatically set as `-xmax`.

```
> ♯ g:Gauss, f:Gauss
> klinfo(distg = 1, paramg = c(0, 1), distf = 1, paramf = c(0.1, 1.5),
xmax = 4)
> klinfo(distg = 1, paramg = c(0, 1), distf = 1, paramf = c(0.1, 1.5),
xmax = 6)
> klinfo(distg = 1, paramg = c(0, 1), distf = 1, paramf = c(0.1, 1.5),
xmax = 8)
```

Table 4.1 shows the numerically obtained K-L information $\hat{I}(g\ f)$ and $\hat{G}(x_k)$, obtained by integrating the density function $g(y)$ from $-x_k$ to x_k, for $x_k = 4, 6$ and 8, and $k = 8, 16, 32$ and 64. It can be seen from Table 4.1 that if x_k (`xmax`) is set sufficiently large, a surprisingly good approximation is obtained even with such small values of k as $k = 16$ or $\Delta x = 0.5$. Note that in this case, we can make sure that the exact value is actually 0.03939922 by equation (4.16). This is because we assume that $g(y)$ follows a normal distribution, and it vanishes to 0 very rapidly as $|x|$ becomes large. When a density function is used for $g(y)$, whose convergence is slower than that of the normal distribution, the accuracy

Table 4.1 *K-L information for various values of x_n and k. (g: normal distribution and f: normal distribution)*

x_k	k	Δx	$\hat{I}(g;f)$	$\hat{G}(x_k)$
4.0	8	1.000	0.03974041	0.99986319
4.0	16	0.500	0.03962097	0.99991550
4.0	32	0.250	0.03958692	0.99993116
4.0	64	0.125	0.03957812	0.99993527
6.0	12	1.000	0.03939929	1.00000000
6.0	24	0.500	0.03939924	1.00000000
6.0	48	0.250	0.03939924	1.00000000
6.0	96	0.125	0.03939923	1.00000000
8.0	16	1.000	0.03939926	1.00000000
8.0	32	0.500	0.03939922	1.00000000
8.0	64	0.250	0.03939922	1.00000000
8.0	128	0.125	0.03939922	1.00000000

of numerical integration can be judged by checking whether $\hat{G}(x_k)$ is close to one.

```
> ♯ g:Gauss, f:Cauchy
> klinfo(distgklinfo(distg = 1, paramg = c(0, 1), distf = 2,
paramf = c(0, 1), xmax = 8)
```

Table 4.2 *Numerical integration for K-L information with various values of k when g(y) is the standard normal distribution and f(y) is a Cauchy distribution.*

x_k	k	Δx	$\hat{I}(g,f)$	$\hat{G}(x_k)$
8.0	16	1.000	0.25620181	1.00000001
8.0	32	0.500	0.25924202	1.00000000
8.0	64	0.250	0.25924453	1.00000000
8.0	128	0.125	0.25924453	1.00000000

Table 4.2 shows the K-L information obtained by the numerical integration when $g(y)$ is the standard normal distribution and $f(y)$ is the standard Cauchy distribution with $\mu = 0$ and $\tau^2 = 1$. It can be seen that

even with a large Δx such as 0.5, we can get very good approximations of $\hat{I}(g;f)$ obtained by using a smaller Δx, and $\hat{G}(x_k)$ is 1 even for $\Delta x = 0.5$.

4.3   Estimation of the K-L Information and the Log-Likelihood

Though the K-L information was introduced as a criterion for the goodness of fit of a statistical model in the previous section, it is rarely used directly to evaluate an actual statistical model except for the case of a Monte Carlo experiment for which the true distribution is known. In an actual situation, we do not know the true distribution $g(y)$, but instead we can obtain the data $y_1, \cdots, y_N$ generated from the true distribution. Hereafter we consider the method of estimating the K-L information of the model $f(y)$ by assuming that the data $y_1, \cdots, y_N$ are independently observed from $g(y)$ (Sakamoto et al. (1986), Konishi and Kitagawa (2008)).

According to the entropy maximization principle, the best model can be obtained by finding the model that maximizes $B(g;f)$ or minimizes $I(g;f)$. As a first step, the K-L information can be decomposed into two terms as

$$I(g;f) = E_Y \log g(Y) - E_Y \log f(Y). \qquad (4.20)$$

Although the first term on the right-hand side of equation (4.20) cannot be computed unless the true distribution $g(y)$ is given, it can be ignored because it is a constant, independent of the model $f(y)$. Therefore, a model that maximizes the second term on the right-hand side is considered to be a good model. This second term is called *expected log-likelihood*. For a continuous model with density function $f(y)$, it is expressible as

$$E_Y \log f(Y) = \int \log f(y) g(y) dy. \qquad (4.21)$$

Unfortunately, the expected log-likelihood still cannot be directly calculated when the true model $g(y)$ is unknown. However, because data y_n is generated according to the density function $g(y)$, due to the *law of large numbers*,

$$\frac{1}{N} \sum_{n=1}^{N} \log f(y_n) \longrightarrow E_Y \log f(Y), \qquad (4.22)$$

holds as the number of data points goes to infinity, i.e., $N \to \infty$.

Therefore, by maximizing the left term, $\sum_{n=1}^{N} \log f(y_n)$, instead of the original criterion $I(g;f)$, we can approximately maximize the entropy. When the observations are obtained independently, N times the term on the left-hand side of (4.22) is called the *log-likelihood*, and it is given by

$$\ell = \sum_{n=1}^{N} \log f(y_n). \tag{4.23}$$

The quantity obtained by taking the exponential of ℓ,

$$L = \prod_{n=1}^{N} f(y_n) \tag{4.24}$$

is called the *likelihood*.

For models used in time series analysis, the assumption that the observations are obtained independently, does not usually hold. For such a general situation, the likelihood is defined by using the joint distribution of $y_1, \cdots, y_N$ as

$$L = f(y_1, \cdots, y_N). \tag{4.25}$$

(4.25) is a natural extension of (4.24), because it reduces to (4.24) when the observations are independently obtained. In this case, the log-likelihood is obtained by

$$\ell = \log L = \log f(y_1, \cdots, y_N). \tag{4.26}$$

4.4 Estimation of Parameters by the Maximum Likelihood Method

If a model contains a parameter θ and its distribution can be expressed as $f(y) = f(y|\theta)$, the log-likelihood ℓ can be considered a function of the parameter θ. Therefore, by expressing the parameter θ explicitly,

$$\ell(\theta) = \begin{cases} \sum_{n=1}^{N} \log f(y_n|\theta), & \text{for independent data} \\ \log f(y_1, \cdots, y_N|\theta), & \text{otherwise} \end{cases} \tag{4.27}$$

is called the *log-likelihood function* of θ.

The log-likelihood function $\ell(\theta)$ evaluates the goodness of the model specified by the parameter θ. Therefore, by selecting θ so as to maximize $\ell(\theta)$, we can determine the optimal value of the parameter of the parametric model, $f(y|\theta)$. The parameter estimation method derived by maximizing the likelihood function or the log-likelihood function is referred to as the *maximum likelihood method*. The parameter estimated by the method of maximum likelihood is called the *maximum likelihood estimate* and is denoted by $\hat{\theta}$.

Under some regularity conditions, the maximum likelihood estimate has the following properties (Huber (1967), Konishi and Kitagawa (2008))

(i) The maximum likelihood estimator $\hat{\theta}$ converges in probabilty to θ_0 as the sample size $N \to \infty$.

(ii) (Central limit theorem) The distribution of $\sqrt{N}(\hat{\theta} - \theta_0)$ converges in law to the normal distribution with the mean vector 0 and the variance covariance matrix $J^{-1}IJ$ as $N \to \infty$, i.e.,

$$\sqrt{N}(\hat{\theta} - \theta_0) \to N(0, J^{-1}IJ), \tag{4.28}$$

where I and J are the *Fisher information matrix* and the negative of the expected Hessian with respect to $\hat{\theta}$ defined by

$$I \equiv E_Y\left[\left\{\frac{\partial}{\partial\theta}\log f(Y|\theta_0)\right\}\left\{\frac{\partial}{\partial\theta}\log f(Y|\theta_0)\right\}^T\right], \tag{4.29}$$

$$J \equiv -E_Y\left[\frac{\partial^2}{\partial\theta\partial\theta^T}\log f(Y|\theta_0)\right]. \tag{4.30}$$

Example (Maximum likelihood estimate of the mean of the normal distribution model)   Consider a normal distribution model with mean μ and variance 1

$$f(y|\mu) = \frac{1}{\sqrt{2\pi}}\exp\left\{-\frac{(y-\mu)^2}{2}\right\}, \tag{4.31}$$

and estimate the mean parameter μ by the maximum likelihood method. Given N observations, $y_1, \cdots, y_N$, the log-likelihood function is given by

$$\ell(\mu) = -\frac{N}{2}\log 2\pi - \frac{1}{2}\sum_{n=1}^{N}(y_n - \mu)^2. \tag{4.32}$$

To maximize the log-likelihood $\ell(\mu)$, it suffices to find μ that minimizes

$$S(\mu) = \sum_{n=1}^{N}(y_n - \mu)^2. \tag{4.33}$$

By equating the first derivative of $S(\mu)$ to zero, we obtain

$$\hat{\mu} = \frac{1}{N}\sum_{n=1}^{N}y_n. \tag{4.34}$$

A method of estimating the parameters of a model by minimizing
the sum of squares such as (4.30) is called the *least squares method*. A
general method of solving the least squares problem will be described
in Chapter 5. For a normal distribution model, the maximum likelihood
estimates of the parameters often coincide with the least squares estimates and can be solved analytically. However, the likelihood or the log-
likelihood functions of a time series model is very complicated in general, and it is not possible to obtain the maximum likelihood estimates or
even their approximate values analytically except for some models such
as the AR model in Chapter 7 and the polynomial trend model in Chapter
11.

In general, the maximum likelihood estimate of the parameter θ
of a time series model is obtained by using a *numerical optimization*
algorithm based on the *quasi-Newton method* described in Appendix A.
According to this method, using the value $\ell(\theta)$ of the log-likelihood and
the first derivative $\partial\ell/\partial\theta$ for a given parameter θ, the maximizer of $\ell(\theta)$
is automatically estimated by repeating

$$\theta_k = \theta_{k-1} + \lambda_k H_{k-1}^{-1} \frac{\partial\ell}{\partial\theta}, \tag{4.35}$$

where θ_0 is an initial estimate of the parameter. The step width λ_k and
the inverse matrix H_{k-1}^{-1} of the Hessian matrix are automatically obtained
by the algorithm.

**Example (Maximum likelihood estimate of the Cauchy distribution
model)**   Assume that ten observations are given as follows.

$$
\begin{array}{ccccc}
-1.10 & -0.40 & -0.20 & -0.02 & 0.02 \\
0.71 & 1.35 & 1.46 & 1.74 & 3.89
\end{array}
$$

The log-likelihood of the Cauchy distribution model

$$f(y|\mu,\tau^2) = \frac{1}{\pi} \frac{\tau}{(y-\mu)^2 + \tau^2,} \tag{4.36}$$

is obtained by

$$\ell(\mu,\tau^2) = 5\log\tau^2 - 10\log\pi - \sum_{n=1}^{10} \log\{(y_n-\mu)^2 + \tau^2\}. \tag{4.37}$$

Table 4.3 *Estimation of the parameters of the Cauchy distribution by the quasi-Newton method.*

k	μ	τ^2	log-likelihood	$\partial\ell/\partial\mu$	$\partial\ell/\partial\tau^2$
0	0.00000	1.00000	−19.1901	2.10968	−0.92404
1	0.38588	0.83098	−18.7140	−0.21335	−0.48655
2	0.34795	0.62966	−18.6536	−0.35810	0.06627
3	0.26819	0.60826	−18.6396	0.00320	−0.01210
4	0.26752	0.60521	−18.6395	0.00000	−0.00002
5	0.26752	0.60520	−18.6395	0.00000	0.00000

The first derivatives of the log-likelihood with respect to the parameter $\theta = (\mu, \tau^2)^T$ are given by

$$\frac{\partial\ell}{\partial\mu} = 2\sum_{n=1}^{10} \frac{y_n - \mu}{(y_n - \mu)^2 + \tau^2} \qquad (4.38)$$

$$\frac{\partial\ell}{\partial\tau^2} = \frac{5}{\tau^2} - \sum_{n=1}^{10} \frac{1}{(y_n - \mu)^2 + \tau^2}. \qquad (4.39)$$

Table 4.3 summarizes the optimization process for obtaining the maximum likelihood estimate of the parameters of the Cauchy distribution when the initial vector is set to $\theta_0 = (0, 1)^T$. The absolute values of $\partial\ell/\partial\mu$ and $\partial\ell/\partial\tau^2$ decrease rapidly, and the maximum likelihood estimate is obtained in five iterations.

As noted in the above example, when the log-likelihood and the first derivatives are obtained analytically, the maximum likelihood estimates of the parameters of the model can be obtained by using a numerical optimization method. However, in time series modeling, it is often difficult to obtain the first derivative of the log-likelihood, because for many time series models, the log-likelihood function is in a very complicated form. For example, for many time series models, the log-likelihood is evaluated numerically, using a Kalman filter. Even in such cases, the maximum likelihood estimate can be obtained by using the first derivative computed by numerical differentiation of the log-likelihood. Table 4.4 shows the result of optimization by this method using the log-likelihood only. The results are almost identical to Table 4.3, and the iteration terminates in a smaller number than the case of the previous table.

Table 4.4 *Estimation of the parameters of the Cauchy distribution by a quasi-Newton method that uses numerical differentiation.*

k	μ	τ^2	log-likelihood	$\partial\ell/\partial\mu$	$\partial\ell/\partial\tau^2$
0	0.00000	1.00000	-19.1901	2.10967	-0.92404
1	0.38588	0.83098	-18.7140	-0.21335	-0.48655
2	0.34795	0.62966	-18.6536	-0.35810	0.06627
3	0.26819	0.60826	-18.6396	0.00320	-0.01210
4	0.26752	0.60521	-18.6395	0.00000	-0.00000

4.5   AIC (Akaike Information Criterion)

So far we have seen that the log-likelihood is a natural estimator of the expected log-likelihood and that the maximum likelihood method can be used for estimation of the parameters of the model. Similarly, if there are several candidate parametric models, it seems natural to estimate the parameters by the maximum likelihood method, and then find the best model by comparing the values of the maximum log-likelihood $\ell(\hat{\theta})$. However, in actuality, the maximum log-likelihood is not directly available for comparisons among several parametric models whose parameters are estimated by the maximum likelihood method, because of the presence of the bias. That is, for the model with the maximum likelihood estimate $\hat{\theta}$, the maximum log-likelihood $N^{-1}\ell(\hat{\theta})$ has a positive bias as an estimator of $E_Y \log f(Y|\hat{\theta})$ (see Figure 4.3 and Konishi and Kitagawa (2008)).

This bias is caused by using the same data twice for the estimation of the parameters of the model and also for the estimation of the expected log-likelihood for evaluation of the model.

The bias of $N^{-1}\ell(\hat{\theta}) \equiv N^{-1}\sum_{n=1}^{N} \log f(y_n|\hat{\theta})$ as an estimate of $E_Y \log f(Y|\hat{\theta})$ is given by

$$C \equiv E_X \left[E_Y \log f(Y|\hat{\theta}) - N^{-1} \sum_{n=1}^{N} \log f(y_n|\hat{\theta}) \right]. \qquad (4.40)$$

Note here that the maximum likelihood estimate $\hat{\theta}$ depends on the sample X and can be expressed as $\hat{\theta}(X)$, and the expectation E_X is taken with respect to X.

Then, correcting the maximum log-likelihood $\ell(\hat{\theta})$ for the bias C, $N^{-1}\ell(\hat{\theta}) + C$ becomes an unbiased estimate of the expected log-likelihood $E_Y \log f(Y|\hat{\theta})$. Here, as will be shown later, since the bias

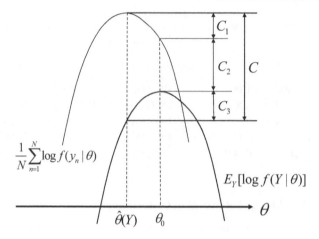

Figure 4.3 *Difference between the expected log-likelihood and the log-likelihood.*

is evaluated as $C = -N^{-1}k$, we obtain the *Akaike Information Criterion* (AIC)

$$
\begin{aligned}
\text{AIC} &= -2\ell(\hat{\theta}) + 2k \\
&= -2\,(\text{maximum log-likelihood}) \\
&\quad + 2\,(\text{number of parameters}).
\end{aligned} \tag{4.41}
$$

In this book, AIC is used as a criterion for model selection (Akaike (1973,1974), Sakamoto et al. (1986), Konishi and Kitagawa (2008)).

Hereinafter, a brief derivation of the AIC will be shown in this section. Readers who are not interested in the model selection criterion itself may skip this part. For more details and other criteria such as BIC and GIC, the readers are referred to Konishi and Kitagawa (2008).

Here it is assumed that the true distribution is $g(y)$, the model distribution is $f(y|\theta)$ and the maximum likelihood estimate of the parameter θ based on data $X = (x_1, \cdots, x_N)$ is denoted by $\hat{\theta} \equiv \hat{\theta}(X)$. On the other hand, the parameter θ_0 that maximizes the expected log-likelihood $E_Y \log f(Y|\theta)$ is called the true parameter. Then, θ_0 satisfies

$$
\frac{\partial}{\partial \theta} E_Y \log f(Y|\theta_0) = 0.
$$

On the other hand, since $\hat{\theta}$ maximizes the log-likelihood function $\ell(\theta) = \sum_{n=1}^{N} \log f(x_n|\theta)$, the following equation holds:

$$\frac{\partial}{\partial \theta} \sum_{n=1}^{N} \log f(x_n|\hat{\theta}) = 0.$$

Here, the terms in (4.37) can be decomposed into three terms (see Figure 4.3).

$$
\begin{aligned}
C &= E_X \left[E_Y \log f(Y|\hat{\theta}) - E_Y \log f(Y|\theta_0) \right] \\
&\quad + E_X \left[E_Y \log f(Y|\theta_0) - N^{-1} \sum_{n=1}^{N} \log f(x_n|\theta_0) \right] \\
&\quad + E_X \left[N^{-1} \sum_{n=1}^{N} \log f(x_n|\theta_0) - N^{-1} \sum_{n=1}^{N} \log f(x_n|\hat{\theta}) \right] \\
&\equiv C_1 + C_2 + C_3.
\end{aligned}
\tag{4.42}
$$

4.5.1 Evaluation of C_1

Consider the Taylor series expansion around θ_0 of the expected log-likelihood $E_Y \log f(Y|\hat{\theta})$ of the model specified by the maximum likelihood estimate up to the second order. Exchanging the order of the differentiation and the expectation, we have

$$
\begin{aligned}
E_Y \log f(Y|\hat{\theta}) &\approx E_Y \log f(Y|\theta_0) + \left\{ \frac{\partial}{\partial \theta} E_Y \log f(Y|\theta_0) \right\} (\hat{\theta} - \theta_0) \\
&\quad + \frac{1}{2} (\hat{\theta} - \theta_0)^T \left\{ \frac{\partial^2}{\partial \theta \partial \theta^T} E_Y \log f(Y|\theta_0) \right\} (\hat{\theta} - \theta_0) \\
&= E_Y \log f(Y|\theta_0) - \frac{1}{2} (\hat{\theta} - \theta_0)^T J (\hat{\theta} - \theta_0).
\end{aligned}
$$

Assume that I and J are the Fisher information matrix and the negative of the expected Hessian defined (4.29) and (4.30), respectively.

Then, according to the central limit theorem (4.28), $\sqrt{N} \left((\hat{\theta} - \theta_0) \right)$ follows a normal distribution with mean 0 and the variance covariance matrix $J^{-1} I J^{-1}$. Therefore, by taking the expectation with respect to X, it follows that

$$
E_X \left[(\hat{\theta} - \theta_0)^T J (\hat{\theta} - \theta_0) \right] = \frac{1}{N} \text{trace} \left\{ I J^{-1} \right\} \approx \frac{k}{N},
\tag{4.43}
$$

where k is the dimension of the matrix I. Note that, if there exists θ_0 such that $g(y) = f(y|\theta_0)$, it follows that $J = I$, and trace $\{IJ^{-1}\} = k$ (Konishi and Kitagawa (2008)). Thus we have an approximation to C_1 :

$$C_1 \equiv E_X\left[E_Y \log f(Y|\hat{\theta}) - E_Y \log f(Y|\theta_0)\right] \approx -\frac{k}{2N}. \tag{4.44}$$

4.5.2 Evaluation of C_3

By the Taylor series expansion of $N^{-1}\sum_{n=1}^{N} \log f(x_n|\theta_0)$ around $\hat{\theta}$, it follows that

$$\frac{1}{N}\sum_{n=1}^{N} \log f(x_n|\theta_0)$$

$$\approx \frac{1}{N}\sum_{n=1}^{N} \log f(x_n|\hat{\theta}) + \frac{1}{N}\sum_{n=1}^{N} \frac{\partial}{\partial\theta} \log f(x_n|\hat{\theta})(\theta_0 - \hat{\theta})$$

$$+ \frac{1}{2}(\theta_0 - \hat{\theta})^T\left\{\frac{1}{N}\sum_{n=1}^{N} \frac{\partial^2}{\partial\theta\partial\theta^T} \log f(x_n|\hat{\theta})\right\}(\theta_0 - \hat{\theta}). \tag{4.45}$$

Since $\hat{\theta}$ is the maximum likelihood estimate, the second term on the right-hand side becomes 0. Moreover, according to the law of large numbers, if $N \to \infty$

$$\frac{1}{N}\sum_{n=1}^{N} \frac{\partial^2}{\partial\theta\partial\theta^T} \log f(x_n|\hat{\theta}) \longrightarrow E_Y\left\{\frac{\partial^2}{\partial\theta\partial\theta^T} \log f(Y|\theta_0)\right\} = -J, \tag{4.46}$$

and we have

$$\frac{1}{N}\sum_{n=1}^{N} \log f(x_n|\theta_0) \approx \frac{1}{N}\sum_{n=1}^{N} \log f(x_n|\hat{\theta}) - \frac{1}{2}(\theta_0 - \hat{\theta})^T J(\theta_0 - \hat{\theta}).$$

Therefore, similarly to (4.44), by taking the expectation of both sides, we have the approximation

$$C_3 = E_X\left[\frac{1}{N}\sum_{n=1}^{N} \log f(x_n|\theta_0) - \frac{1}{N}\sum_{n=1}^{N} \log f(x_n|\hat{\theta})\right] \approx -\frac{k}{2N}. \tag{4.47}$$

4.5.3 Evaluation of C_2

Since the expectation of $\log f(x_n|\theta_0)$ becomes the expected log-likelihood for any fixed θ_0, we have

$$C_2 = E_X\left[E_Y \log f(Y|\theta_0) - \frac{1}{N}\sum_{n=1}^{N} \log f(x_n|\theta_0)\right] = 0. \tag{4.48}$$

4.5.4   Evaluation of C and AIC

By summing up the three expressions (4.44), (4.47) and (4.48), we have the approximation

$$C = \mathrm{E}_X \left[\mathrm{E}_Y \log f(Y|\hat{\theta}) - \frac{1}{N} \sum_{n=1}^{N} \log f(x_n|\hat{\theta}) \right] \approx -\frac{k}{N}, \qquad (4.49)$$

and this shows that $N^{-1}\ell(\hat{\theta})$ is larger than $\mathrm{E}_Y \log f(Y|\hat{\theta})$ by k/N on average.

Therefore, it can be seen that

$$N^{-1}\ell(\hat{\theta}) + C \approx N^{-1}\left(\ell(\hat{\theta}) - k\right) \qquad (4.50)$$

is an approximately unbiased estimator of the expected log-likelihood $\mathrm{E}_Y \log f(Y|\hat{\theta})$ of the maximum likelihood model. The Akaike information criterion (AIC) is defined by multiplying (4.46) by $-2N$, i.e.

$$
\begin{aligned}
\mathrm{AIC} &= -2\ell(\hat{\theta}) + 2k \\
&= -2\,(\text{maximum log-likelihood}) \\
&\quad + 2\,(\text{number of parameters}). \qquad (4.51)
\end{aligned}
$$

Since minimizing the AIC is approximately equivalent to minimizing the K-L information, an approximately optimal model is obtained by selecting the model that minimizes AIC. A reasonable and automatic model selection thus becomes possible by using the AIC.

4.6   Transformation of Data

When we look at graphs of positive valued processes such as the number of occurrences of a certain event, the number of people or the amount of sales, we would find that the variance increases together with an increase of the mean value, and the distribution is highly skewed. It is difficult to analyze such data using a simple linear Gaussian time series model since the characteristics of the data change with its level and the distribution is significantly different from the normal distribution. For such a case we shall use the nonstationary models in Chapters 11, 12 and 13, or the non-Gaussian models in Chapters 14 and 15.

```
> par(mar=c(2,2,2,1)+0.1)
> data(WHARD)
> plot(log10(WHARD))
> data(Sunspot)
> plot(log10(Sunspot))
```

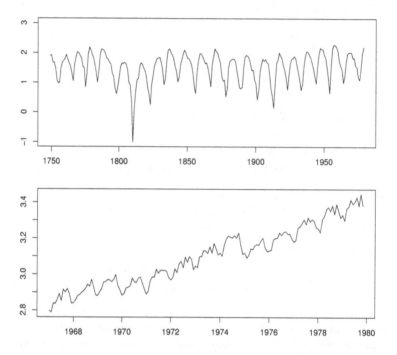

Figure 4.4 *Log-transformation of the sunspot number data and the WHARD data.*

Even in those cases, the variance of the log-transformed series $z_n = \log y_n$ might become almost uniform, and its marginal distribution might be reasonably approximated by a normal distribution. In Figure 4.4, it can be seen that the highly skewed sunspot number data shown in plot (b) of Figure 1.1 becomes almost symmetric after log-transformation. Furthermore, although the variance of the WHARD data shown in plot (e) of Figure 1.1 increases with time, after the log-transformation the variance of the series becomes almost a constant.

The *Box-Cox transformation* is well known as a generic data transformation (Box and Cox (1964)), which includes the log-transformation as a special case

$$
z_n = \begin{cases} \lambda^{-1}(y_n^{\lambda} - 1), & \text{if } \lambda \neq 0 \\ \log y_n, & \text{if } \lambda = 0. \end{cases}
\tag{4.52}
$$

Ignoring a constant term, the Box-Cox transformation yields the logarithm of the original series for $\lambda = 0$, the inverse for $\lambda = -1$ and the square root for $\lambda = 0.5$. In addition, it agrees with the original data for $\lambda = 1$.

Applying the information criterion AIC to the Box-Cox transformation, we can determine the best parameter λ of the Box-Cox transformation (Konishi and Kitagawa (2008)). On the assumption that the density function of the data $z_n = h(y_n)$ obtained by the Box-Cox transformation of data y_n is given by $f(z)$, then the density function of y_n is obtained by

$$g(y) = \left| \frac{dh}{dy} \right| f(h(y)), \qquad (4.53)$$

here, $|dh/dy|$ is called the *Jacobian* of the transformation. The equation (4.53) implies that a model for the transformed series automatically determines a model for the original data. For instance, assume that the values of AIC's of the normal distribution models fitted to the original data y_n and transformed data z_n are evaluated as AIC_y and AIC_z, respectively. Then, it can be judged which, the original data or the transformed data, is closer to a normal distribution by comparing the values of AIC_y and

$$\text{AIC}'_z = \text{AIC}_z - 2 \sum_{i=1}^{N} \log \left| \frac{dh}{dy} \right|_{y=y_i}. \qquad (4.54)$$

Namely, it will be considered that the original data is better than the transformed data, if $\text{AIC}_y < \text{AIC}'_z$, and the transformed data is better, if $\text{AIC}_y > \text{AIC}'_z$. Further, by finding a value that minimizes AIC'_z, the optimal value λ of the Box-Cox transformation can be selected. However, in actual time series modeling, we usually fit various time series models to the Box-Cox transformation of the data. Therefore, in such a situation, it is necessary to correct the AIC of the time series model with the Jacobian of the Box-Cox transformation.

Example (Box-Cox transformation of the sunspot number data)

```
> par(mar=c(2,2,2,1)+0.1)
> data(Sunspot)
> boxcox(Sunspot)
```

The Box-Cox transformation of data and the AIC of the Gaussian model fitted to the transformed data evaluated as the model for the original data can be obtained by the function boxcox of the TSSS package. Table 4.5 shows the summary of the Box-Cox transformations with various values of λ to the sunspot number data of Figure 1.1(b). In

Table 4.5 *Various Box-Cox transforms and their AIC's for sunspot number data.*

λ	AIC'	log-L	AIC	log-L	mean	variance
1.0	2360.37	-1178.19	2360.37	-1178.19	49.109	1576.35
0.8	2313.88	-1154.94	1992.09	-994.04	25.917	320.08
0.6	2281.75	-1138.87	1638.16	-817.08	14.382	69.16
0.4	2267.00	-1131.50	1301.62	-648.81	8.442	16.11
0.2	2274.40	-1135.20	987.23	-491.61	5.261	4.13
0.0	2313.40	-1154.70	704.44	-350.22	3.483	1.21
-0.2	2405.33	-1200.67	474.58	-235.29	2.441	0.45
-0.4	2587.43	-1291.71	334.88	-165.44	1.800	0.25
-0.6	2881.56	-1438.78	307.22	-151.61	1.386	0.22
-0.8	3260.47	-1628.23	364.33	-180.17	1.103	0.28
-1.0	3685.11	-1840.56	467.18	-231.59	8.996	0.43

this case, it can be seen that the transformation with $\lambda = 0.4$, that is, $z_n = 2.5(y_n^{0.4} - 1)$ is the best Box-Cox transformation, and the AIC of the transformation is 2267.00.

Figure 4.5 shows the original sunspot data and its best Box-Cox transformation. It can be seen that by this transformation, the original skewed time series was transformed to a reasonably symmetrized time series.

```
> data(WHARD)
> boxcox(WHARD)
```

Similarly, Table 4.6 shows the summary of Box-Cox transformations for the WHARD data. In this case, $\lambda = -0.3$ attains the minimum AIC, 2355.33. Figure 4.6 shows the original WHARD data and the Box-Cox transformed data obtained by $\hat{\lambda} = -0.3$. It can be seen that the amplitude of the fluctuation around the trend was approximately equalized by this best Box-Cox transformation.

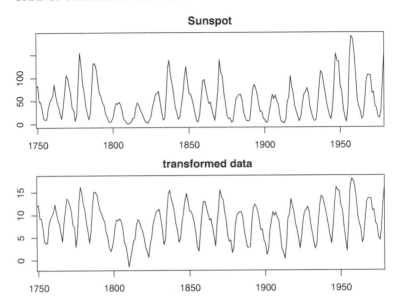

Figure 4.5: *Sunspot number data and the best Box-Cox transformed data.*

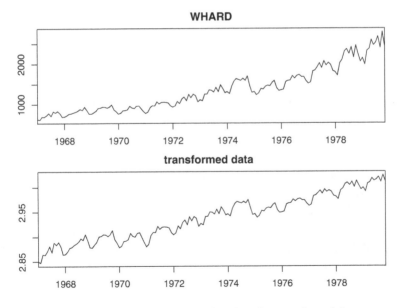

Figure 4.6: *WHARD data and the best Box-Cox transformed data.*

Table 4.6: *Various Box-Cox transforms and their AIC's for WHARD data*

λ	AIC'	log-L	AIC	log-L	mean	variance
1.0	2387.94	-1191.97	2387.94	-1191.97	1348.090	2.80e+05
0.8	2378.55	-1187.27	1936.27	-966.14	393.043	1.52e+04
0.6	2370.74	-1183.37	1486.19	-741.09	121.978	8.33e+02
0.4	2364.52	-1180.26	1037.70	-516.85	41.379	4.61e+01
0.2	2359.90	-1177.95	590.81	-293.40	15.886	2.58e+00
0.0	2356.89	-1176.44	145.51	-70.76	7.133	1.46e-01
-0.2	2355.45	-1175.73	-298.19	151.10	3.796	8.33e-03
-0.3	2355.33	-1175.66	-519.46	261.73	2.939	2.00e-03
-0.4	2355.59	-1175.79	-740.33	372.17	2.354	4.81e-04
-0.6	2357.27	-1176.63	-1180.93	592.46	1.644	2.80e-05
-0.8	2360.46	-1178.23	-1620.01	812.01	1.246	1.65e-06
-1.0	2365.12	-1180.56	-2057.62	1030.81	0.999	9.80e-08

Problems

1. Given n integers $\{m_1,\ldots,m_n\}$, obtain the maximum likelihood estimate $\hat{\lambda}$ of the Poisson distribution model $f(m|\lambda) = e^{-\lambda}\lambda^m/m!$.

2. Given two sets of data $\{x_1,\ldots,x_n\}$ and $\{y_1,\ldots,y_m\}$ that follow normal distributions;

   (1) Assuming that the variances are the same, show a method of checking whether or not the means are identical by using the AIC.

   (2) Assuming that the means are the same, show a method of checking whether or not the variances are identical by using the AIC.

3. Given data $\{y_1,\ldots,y_n\}$, consider a method of deciding whether a Gaussian model or a Cauchy model is better by using the AIC.

4. In tossing a coin n times, a head occurred m times.

   (1) Obtain the probability of the occurrence of a head.

   (2) Consider a method of deciding whether or not this is a fair coin, based on the AIC.

5. Assume that the true density function is $g(y)$ and the model is $f(y|\theta)$. If there exists θ_0 such that $g(y) = f(y|\theta_0)$, show that $J = I$.

6. Obtain the density function of y when $z = \log y$ follows a Gaussian distribution $N(\mu,\sigma^2)$.

Chapter 5

The Least Squares Method

For many regression models and time series models that assume normality of the noise distribution, least squares estimates often coincide with or provide good approximations to the maximum likelihood estimates of the unknown parameters. This chapter explains the Householder transformation as a convenient method to obtain least squares estimates of regression models (Golub (1965), Sakamoto et al. (1986)). With this method, we can obtain precise estimates of the coefficients of the model and perform order selection or variable selection based on the information criterion AIC quite efficiently.

5.1 Regression Models and the Least Squares Method

Given the objective variable, y_n, and the m explanatory variables, $x_{n1}, \cdots, x_{nm}$, a model that expresses the variation of y_n by the linear combination of the explanatory variables

$$y_n = \sum_{i=1}^{m} a_i x_{ni} + \varepsilon_n \tag{5.1}$$

is called a *regression model*. Here, a_i is called the *regression coefficient* of the explanatory variable x_{ni} and the number of explanatory variables m is referred to as the *order* of the model. Moreover, the difference between the objective variable y_n and the predicted value based on the explanatory variables, ε_n, is called the residual, and is assumed to be an independent random variable that follows a normal distribution with mean 0 and variance σ^2.

Define the N-dimensional vector y and the $N \times m$ matrix Z as

$$y = \begin{bmatrix} y_1 \\ y_2 \\ \vdots \\ y_N \end{bmatrix}, \quad Z = \begin{bmatrix} x_{11} & \cdots & x_{1m} \\ x_{21} & \cdots & x_{2m} \\ \vdots & & \vdots \\ x_{N1} & \cdots & x_{Nm} \end{bmatrix}, \tag{5.2}$$

then the regression model can be concisely expressed by the matrix-vector representation

$$y = Za + \varepsilon. \tag{5.3}$$

The vector y and the matrix Z are called the vector of objective variables and the matrix of explanatory variables (or design matrix), respectively. $a = (a_1, \cdots, a_m)^T$ is a vector of regression coefficients, and $\varepsilon = (\varepsilon_1, \cdots, \varepsilon_N)^T$ is a vector of residuals. The regression model (5.1) contains the regression coefficients $a_1, \cdots, a_m$, and the variance σ^2 as the parameters, and we can express these as a vector of parameters, $\theta = (a_1, \cdots, a_m, \sigma^2)^T$.

When N independent observations $\{y_n, x_{n1}, \cdots, x_{nm}\}$, $n = 1, \cdots, N$, are given, the likelihood and log-likelihood of the regression model become functions of θ and are given by

$$L(\theta) = \prod_{n=1}^{N} p(y_n | \theta, x_{n1}, \cdots, x_{nm}) \tag{5.4}$$

$$\ell(\theta) = \sum_{n=1}^{N} \log p(y_n | \theta, x_{n1}, \cdots, x_{nm}), \tag{5.5}$$

respectively. Here, from equation (5.1), each term of the right-hand side of the above equations can be expressed as

$$p(y_n | \theta, x_{n1}, \cdots, x_{nm}) = \frac{1}{\sqrt{2\pi\sigma^2}} \exp\left\{ -\frac{1}{2\sigma^2} \left(y_n - \sum_{i=1}^{m} a_i x_{ni} \right)^2 \right\}, \tag{5.6}$$

$$\log p(y_n | \theta, x_{n1}, \cdots, x_{nm}) = -\frac{1}{2} \log 2\pi\sigma^2 - \frac{1}{2\sigma^2} \left(y_n - \sum_{i=1}^{m} a_i x_{ni} \right)^2, \tag{5.7}$$

and the log-likelihood function is given by

$$\ell(\theta) = -\frac{N}{2} \log 2\pi\sigma^2 - \frac{1}{2\sigma^2} \sum_{n=1}^{N} \left(y_n - \sum_{i=1}^{m} a_i x_{ni} \right)^2. \tag{5.8}$$

The maximum likelihood estimate $\hat{\theta} = (\hat{a}_1, \cdots, \hat{a}_m, \hat{\sigma}^2)^T$ of the parameter θ can be obtained by finding the value of θ that maximizes the log-likelihood function $\ell(\theta)$. Given any set of regression coefficients $a_1, \cdots, a_m$, the maximum likelihood estimate of σ^2 can be obtained by solving the normal equation

$$\frac{\partial \ell(\theta)}{\partial \sigma^2} = -\frac{N}{2\sigma^2} + \frac{1}{2(\sigma^2)^2} \sum_{n=1}^{N} \left(y_n - \sum_{i=1}^{m} a_i x_{ni} \right)^2 = 0. \tag{5.9}$$

Therefore, σ^2 can be easily obtained as

$$\hat{\sigma}^2 = \frac{1}{N} \sum_{n=1}^{N} \left(y_n - \sum_{i=1}^{m} a_i x_{ni} \right)^2. \tag{5.10}$$

Then, substituting this into (5.8), the log-likelihood becomes a function of the regression coefficients $a_1, \cdots, a_m$ and is given by

$$\ell(a_1, \cdots, a_m) = -\frac{N}{2} \log 2\pi \hat{\sigma}^2 - \frac{N}{2}. \tag{5.11}$$

Since the logarithm is a monotone increasing function, the regression coefficients $a_1, \cdots, a_m$ that maximize the log-likelihood (5.11) are obtained by minimizing the variance $\hat{\sigma}^2$ in (5.10). Therefore, it can be seen that the maximum likelihood estimates of the parameters of the regression model are obtained by the least squares method.

5.2 The Least Squares Method Based on the Householder Transformation

As shown in the previous section, the maximum likelihood estimates of the regression coefficients of linear regression models can be obtained by the *least squares method* that minimizes (5.10). Here, using the matrix-vector notation of (5.2) and (5.3), the residual sum of squares can be simply expressed as

$$\sum_{n=1}^{N} \left(y_n - \sum_{i=1}^{m} a_i x_{ni} \right)^2 = \|y - Za\|_N^2 = \|\varepsilon\|_N^2, \tag{5.12}$$

where $\|y\|_N$ denotes the Euclidean norm of the N-dimensional vector y. The well-known derivation of the least squares method is to set the partial derivative of $\|y - Za\|_N^2$ in (5.12) with respect to the parameter a equal to zero, which results in the normal equation $Z^T y = Z^T Z a$. Then solving this equation, we obtain the least squares estimates of a by $\hat{a} = (Z^T Z)^{-1} Z^T y$.

However, in actual computation, it is convenient to use the following method based on orthogonal transformation, since it can yield accurate estimates, and it is suited for various kinds of modeling operations (Golub (1965), Sakamoto et al. (1986)).

For any $N \times N$ orthogonal matrix U, the norm of the vector $y - Za$ is unchanged, even if it is transformed by U. Namely, we have that

$$\|y - Za\|_N^2 = \|U(y - Za)\|_N^2 = \|Uy - UZa\|_N^2, \tag{5.13}$$

and this implies that the vector a that minimizes $\|Uy - UZa\|_N^2$ is identical to the one that minimizes $\|y - Za\|_N^2$. Therefore, to obtain the least squares estimates of a, we can first apply an orthogonal transformation to make UZ an adequate form and then find a vector a that minimizes (5.13).

The least squares method based on (5.13) can be realized very efficiently by using a *Householder transformation* as follows (Golub (1965), Sakamoto et al. (1986)). First, define an $N \times (m+1)$ matrix

$$X = [Z|y], \tag{5.14}$$

by augmenting with the vector of objective variables y to the right of the matrix of the explanatory variables Z.

Applying a suitable Householder transformation U to the matrix X, it can be transformed to an upper triangular matrix S as

$$UX = S = \begin{bmatrix} s_{11} & \cdots & s_{1m} & s_{1,m+1} \\ & \ddots & \vdots & \vdots \\ & & s_{mm} & s_{m,m+1} \\ & & & s_{m+1,m+1} \\ & O & & \end{bmatrix}. \tag{5.15}$$

Here, since the first m rows and the $(m+1)$-th rows of S correspond to UZ and Uy in equation (5.13), respectively, we have that

$$\|Uy - UZa\|_N^2 = \left\| \begin{bmatrix} s_{1,m+1} \\ \vdots \\ s_{m,m+1} \\ s_{m+1,m+1} \\ 0 \\ \vdots \\ 0 \end{bmatrix} - \begin{bmatrix} s_{11} & \cdots & s_{1m} \\ & \ddots & \vdots \\ & & s_{mm} \\ & O & \end{bmatrix} \begin{bmatrix} a_1 \\ \vdots \\ a_m \end{bmatrix} \right\|_N^2$$

$$= \left\| \begin{bmatrix} s_{1,m+1} \\ \vdots \\ s_{m,m+1} \end{bmatrix} - \begin{bmatrix} s_{11} & \cdots & s_{1m} \\ & \ddots & \vdots \\ O & & s_{mm} \end{bmatrix} \begin{bmatrix} a_1 \\ \vdots \\ a_m \end{bmatrix} \right\|_m^2 + s_{m+1,m+1}^2 \tag{5.16}$$

It should be noted that the second term $s_{m+1,m+1}^2$ on the right-hand side of (5.16) does not depend on the value of a and takes a constant value. Therefore, the least squares estimate is obtained by finding the vector $a = (a_1, \cdots, a_m)^T$ that attains the minimum of the first term, namely 0. This shows that the least squares estimate of a is obtained as the solution to the linear equation

$$
\begin{bmatrix} s_{11} & \cdots & s_{1m} \\ & \ddots & \vdots \\ O & & s_{mm} \end{bmatrix} \begin{bmatrix} a_1 \\ \vdots \\ a_m \end{bmatrix} = \begin{bmatrix} s_{1,m+1} \\ \vdots \\ s_{m,m+1} \end{bmatrix}. \tag{5.17}
$$

The linear equation (5.17) can be easily solved by backward substitution, because the matrix on the left-hand side is in upper triangular form. That is, we can obtain $a = (a_1, \cdots, a_m)^T$ by

$$
\hat{a}_m = \frac{s_{m,m+1}}{s_{mm}} \tag{5.18}
$$

$$
\hat{a}_i = \frac{(s_{i,m+1} - s_{i,i+1}\hat{a}_{i+1} - \cdots - s_{i,m}\hat{a}_m)}{s_{ii}}, \quad i = m-1, \cdots, 1.
$$

Further, since $s_{m+1,m+1}^2$ is the sum of squares of the residual vector, the least squares estimate of the residual variance σ^2 of the regression model with order m is obtained by

$$
\hat{\sigma}_m^2 = \frac{s_{m+1,m+1}^2}{N}. \tag{5.19}
$$

5.3   Selection of Order by AIC

By substituting the estimate of the residual variance obtained from (5.19) into (5.11), the maximum log-likelihood becomes

$$
\ell(\hat{\theta}) = -\frac{N}{2}\log\left(2\pi\hat{\sigma}_m^2\right) - \frac{N}{2}. \tag{5.20}
$$

The regression model with order m has $m+1$ parameters, $a_1, \cdots, a_m$ and σ^2. Therefore, the AIC of the regression model with order m is given by

$$
\begin{aligned}
\text{AIC}_m &= -2\ell(\hat{\theta}) + 2\,(\text{number of parameters}) \\
&= N(\log\left(2\pi\hat{\sigma}_m^2\right) + 1) + 2\,(m+1). \tag{5.21}
\end{aligned}
$$

If the upper triangular matrix S in (5.15) is given, not only the regression model with order m, but also all regression models with order less than m can be obtained. That is, for $j \leq m$, the estimate of the residual variance and the AIC of the regression model with j explanatory variables, $x_{n1}, \cdots, x_{nj}$,

$$
y_n = \sum_{i=1}^{j} a_i x_{ni} + \varepsilon_n, \tag{5.22}
$$

are obtained by (Sakamoto et al. (1986))

$$\hat{\sigma}_j^2 = \frac{1}{N} \sum_{i=j+1}^{m+1} s_{i,m+1}^2$$

$$\text{AIC}_j = N(\log 2\pi \hat{\sigma}_j^2 + 1) + 2(j+1). \tag{5.23}$$

The least squares estimates of the regression coefficients can be obtained by solving the linear equation by backward substitution,

$$\begin{bmatrix} s_{11} & \cdots & s_{1j} \\ & \ddots & \vdots \\ & & s_{jj} \end{bmatrix} \begin{bmatrix} a_1 \\ \vdots \\ a_j \end{bmatrix} = \begin{bmatrix} s_{1,m+1} \\ \vdots \\ s_{j,m+1} \end{bmatrix}. \tag{5.24}$$

To perform *order selection* using the AIC, we need to compute $\text{AIC}_0, \cdots, \text{AIC}_m$ by (5.23), and to look for the order that achieves the smallest value. Here we note that once the upper triangular matrix S is obtained, the AIC's of the regression models of all orders can be immediately computed by (5.23) without estimating the regression coefficients. Therefore, the estimation of the regression coefficients by (5.24) is necessary only for the model with the order that attains the minimum value of the AIC.

Example (Trigonometric regression model)

The function lsqr of the package TSSS fits the trigonometric regression model by the least squares method based on Housholder transformation. The maximum order of the regression model k can be specified by the parameter lag. If the lag is not specified, the default value is $\sqrt{N}$, where N is the datalength.

```
> data(Temperature)
> lsqr(Temperature, lag=10)
> lsqr(Temperature)
```

Table 5.1 summarizes the residual variances and the AIC's when trigonometric regression models with m up to 10

$$y_n = a + \sum_{j=1}^{m} b_j \sin(j\omega n) + \sum_{j=1}^{\ell} c_j \cos(j\omega n) + \varepsilon_n \tag{5.25}$$

are fitted to the maximum temperature data shown in Figure 1.1(c). Here, ℓ is either m or $m-1$. The numbers in the right-most column

Table 5.1 *Residual variances and AIC's of regression models of various orders.*
p: number of regression coefficients of the model, σ^2: residual variance, differ-
ence of AIC from its minimum.

Order	σ^2	AIC	AIC-AICMIN
0	4.026179e+02	4296.230	1861.168
1	6.008618e+01	3373.757	938.694
2	4.453795e+01	3230.230	795.168
3	9.291295e+00	2470.540	35.477
4	9.286903e+00	2472.310	37.248
5	9.283630e+00	2474.139	39.076
6	9.279759e+00	2475.936	40.874
7	9.266183e+00	2477.225	42.162
8	9.266175e+00	2479.224	44.162
9	9.263455e+00	2481.082	46.019
10	9.038053e+00	2471.110	36.047
11	9.037097e+00	2473.059	37.996
12	8.708213e+00	2457.042	21.979
13	8.704562e+00	2458.838	23.775
14	8.642518e+00	2457.362	22.299
15	8.636084e+00	2459.000	23.937
16	8.420505e+00	2448.714	13.651
17	8.400768e+00	2449.573	14.511
18	8.236523e+00	2441.977	6.915
19	8.099144e+00	2435.803	0.740
20	8.053609e+00	2435.063	0.000
21	8.046706e+00	2436.646	1.583

min AIC = 2435.063 attained at m = 20, σ^2 = 8.053609e+00

in the Table 5.1 show the differences of the AIC's from the minimum
AIC value. The explanatory variables were assumed to be adopted in
the order of $\{1, \sin \omega n, \cos \omega n, \cdots, \sin k\omega n, \cos k\omega n\}$. Therefore, the
parameter vector of the model with the highest order becomes $\theta = (a, b_1, c_1, \cdots, b_k, c_k)^T$. The number of regression coefficients is $p = 2m$
for $\ell = m - 1$ and $p = 2m + 1$ for $\ell = m$. The model with the highest
order $k = 10$ has 21 explanatory variables. Since a strong annual cycle
was seen in this data, it was assumed that $\omega = 2\pi/365$.

The function lsqr computes AIC, σ^2 and regression coefficients for
each order and the MAICE order and the regression curve obtained
by the MAICE model. They are given as the elements aic, sigma2,
regress, maice.order and tripoly. Therefore, for example, if
the graph of the AIC values is necessary, we can obtain as follows:

```
> z <- lsqr(Temperature)
> x <- seq(0,21,length=22)
> par( mfrow = c(1,2) )
> plot(x,z$aic,type="b",pch=20)
> plot(x,z$aic,type="b",pch=20,ylim=c(2430,2490))
```

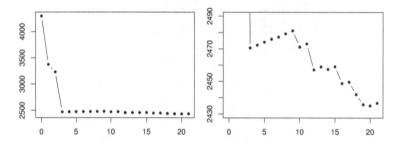

Figure 5.1 *AIC's of the trigonometric regression models for the maximum temperature data.*

Figure 5.1 shows the change of the AIC when the the number of exprlanetary variables were increased up to 21. From Table 5.1 and Figure 5.1, it can be seen that the model with order 20, i.e. the model composed of a constant term, 10 sine components and 9 cosine components attains the minimum of AIC.

Figure 5.2 shows the original data and the regression curve obtained by this model.

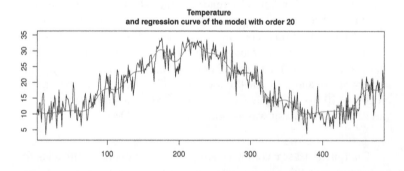

Figure 5.2 *Original data and the minimum AIC regression curve for the maximum temperature data.*

5.4 Addition of Data and Successive Householder Reduction

Utilizing the properties of orthogonal transformations, it is easy to update the model by addition of data (Kitagawa and Akaike (1978)). If we need to fit a regression model to a huge data set and try to save the matrix of (5.14), the computer memory may fill up and we may not be able to run the program.

Even in this case, repeated application of the Householder transformation introduced in this section can yield the upper triangular matrix in (5.13). That is, if our computer has a memory sufficient to store the area of the $L \times (m+1)$ matrix (here, $L > m+1$), then the upper triangular matrix S can be obtained by dividing the data into several sub-data-sets with data length less than or equal to $L - m - 1$.

Assuming that an upper triangular matrix S has already been obtained from N sets of data $\{y_n, x_{n1}, \cdots, x_{nm}\}$, $n = 1, \cdots, N$, we can effectively obtain a regression model from the matrix S, as shown in section (5.15). Here, we assume that M new sets of data $\{y_n, x_{n1}, \cdots, x_{nm}\}$, $n = N + 1, \cdots, N+M$, are obtained. Then, in order to fit a regression model to the entire $N + M$ sets of data, we have to construct the $(N+M) \times (m+1)$ matrix

$$X_1 = \begin{bmatrix} x_{11} & \cdots & x_{1m} & y_1 \\ \vdots & \ddots & \vdots & \vdots \\ x_{N+M,1} & \cdots & x_{N+M,m} & y_{N+M} \end{bmatrix}, \qquad (5.26)$$

instead of (5.14) and then transform this into an upper triangular matrix by a Householder transformation $S' = U'X_1$. Unfortunately, this method cannot utilize the results of the computation for the previous data sets, and we need a large storage for preparing the $(N+M) \times (m+1)$ matrix X_1.

However, since the Householder transformation is an orthogonal transformation, it can be shown that the same matrix as S' can be obtained by building an $(M+m+1) \times (m+1)$ matrix by augmenting an $M \times (m+1)$ matrix under the triangular matrix (5.15),

$$X_2 = \begin{bmatrix} s_{11} & \cdots & s_{1m} & s_{1,m+1} \\ & \ddots & \vdots & \vdots \\ & & s_{mm} & s_{m,m+1} \\ & \mathbf{O} & & s_{m+1,m+1} \\ x_{N+1,1} & \cdots & x_{N+1,m} & y_{N+1} \\ \vdots & \ddots & \vdots & \vdots \\ x_{N+M,1} & \cdots & x_{N+M,m} & y_{N+M} \end{bmatrix}, \qquad (5.27)$$

and reducing it to an upper triangular matrix by a Householder transformation.

Therefore, to perform a Householder transformation of data length longer than L, we first obtain an upper triangular matrix S by putting $N = L$, and then repeat the update of S by adding $M = L - m - 1$ data.

On the other hand, if the upper triangular matrix S_2 has already been obtained from a new data set $\{y_n, x_{n1}, \cdots, x_{nm}\}$, $n = N + 1, \cdots, N + M$, then we define a $2(m+1) \times (m+1)$ matrix as

$$X_3 = \begin{bmatrix} S_1 \\ S_2 \end{bmatrix}, \tag{5.28}$$

and by reducing it to upper triangular form, we can obtain the same matrix as S'.

For $M \gg m$, since the number of rows of the matrix X_3 is smaller than the number of rows of X_1 and X_2, the amount of computation for the Householder transformation of X_3 is significantly less than that required for the transformation of (5.27). This method will be used in Chapter 8 to fit a locally stationary AR model.

5.5 Variable Selection by AIC

In Section 5.3, the method of selecting the order for the model by AIC was explained. However, in that section, it was implicitly assumed that the order of adopting the explanatory variables was provided beforehand and only a model of the form

$$y_n = \sum_{i=1}^{j} a_i x_{ni} + \varepsilon_n \tag{5.29}$$

was considered.

This method of selecting variables is quite natural for the autoregressive model shown in Section 6.1 and the polynomial regression model shown in Section 11.1. However, for a multivariate regression model and multivariate time series models, the order of adopting variables as explanatory variables is not generally provided beforehand. Assuming that $(\ell_1, \cdots, \ell_m)$ is an index vector indicating the order of adopting the explanatory variables, the optimal model could be selected among models of the form

$$y_n = \sum_{i=1}^{j} a_{\ell_i} x_{n,\ell_i} + \varepsilon_n. \tag{5.30}$$

In this case, even if the order j is specified, there are $_mC_j$ different models, depending on the index vector. Such a model is called a *subset regression model*. To fit subset regression models with explanatory variables in the order of $(\ell_1, \cdots, \ell_m)$, transform the upper triangular matrix S to the matrix T that consists of $m+1$ column vectors with the numbers of non-zero elements given by $j_1, \cdots, j_m$, respectively, by the Householder transformation. Here, $j_1, \cdots, j_m$ is the inverse function of the index vector $(\ell_1, \cdots, \ell_m)$ satisfying $\ell_{j_i} = i$.

Example   For the case of $m = 4$ and $(\ell_1, \ell_2, \ell_3, \ell_4) = (2, 4, 3, 1)$, inverse function is $(j_1, \cdots, j_m) = (4, 1, 3, 2)$, and the matrix T is given by

$$
T = \begin{bmatrix}
t_{11} & t_{12} & t_{13} & t_{14} & t_{15} \\
t_{21} & 0 & t_{23} & t_{24} & t_{25} \\
t_{31} & 0 & t_{33} & 0 & t_{35} \\
t_{41} & 0 & 0 & 0 & t_{45} \\
0 & 0 & 0 & 0 & t_{55}
\end{bmatrix}. \tag{5.31}
$$

Then, the residual variance and the AIC of the model that uses the j explanatory variables $\{x_{nl_1}, \cdots, x_{nl_j}\}$ are given by

$$
\begin{aligned}
\hat{\sigma}^2(\ell_1, \cdots, \ell_j) &= \frac{1}{N} \sum_{i=j+1}^{m+1} t_{i,m+1}^2 \\
\mathrm{AIC}(\ell_1, \cdots, \ell_j) &= N \log 2\pi \hat{\sigma}^2(\ell_1, \cdots, \ell_j) + N + 2(j+1).
\end{aligned} \tag{5.32}
$$

Regression coefficients are then obtained by solving the linear equation

$$
\begin{bmatrix}
t_{1,\ell_1} & \cdots & t_{1,\ell_j} \\
 & \ddots & \vdots \\
O & & t_{j,\ell_j}
\end{bmatrix}
\begin{bmatrix}
a_{\ell_1} \\
\vdots \\
a_{\ell_j}
\end{bmatrix}
=
\begin{bmatrix}
t_{1,m+1} \\
\vdots \\
t_{j,m+1}
\end{bmatrix} \tag{5.33}
$$

by backward substitution.

However, in actual computation, it is not necessary to actually exchange the order of explanatory variables and reduce the matrix to upper triangular form. We can easily obtain them from the upper triangular matrix T of (5.31) by the following backward substitution:

$$
\hat{a}_{\ell_j} = t_{j,\ell_j}^{-1} t_{j,m+1} \tag{5.34}
$$

$$
\hat{a}_{\ell_i} = t_{i,\ell_i}^{-1} (t_{i,m+1} - t_{i,\ell_{i+1}} \hat{a}_{\ell_{i+1}} - \cdots - t_{i,\ell_j} \hat{a}_{\ell_j}), \quad i = j-1, \cdots, 1.
$$

Problems

1. Obtain the AIC when the variance σ^2 is known in the regression model in (5.1).

2. Assume that N pairs of data $\{x_1, y_1\}, \ldots, \{x_N, y_N\}$ are given.

   (1) Obtain the least squares estimates $\hat{a}$ and $\hat{b}$ of the second order polynomial regression model $y_n = ax_n^2 + bx_n + \varepsilon_n$ that passes through the origin.

   (2) Obtain a second order polynomial regression model that passes through the origin and the point $(c, 0)$, and consider how to obtain the least squares estimate of the model.

Chapter 6

Analysis of Time Series Using ARMA Models

The characteristics of time series can be concisely described using time series models. Further, the time series model can be used for prediction, signal extraction and decision making for the time series. In this chapter, we consider methods for obtaining the impulse response function, the autocovariance function, the partial autocorrelation (PARCOR), the power spectrum and the roots of the characteristic equation from the univariate ARMA model (Box and Jenkins (1970), Brockwell and Davis (1991), Shumway and Stoffer (2000)). The relations between the AR coefficients and the PARCOR's are also shown. Further, for multivariate time series, the cross-spectrum and the relative power contribution are derived from the multivariate AR model.

6.1 ARMA Model

A model that expresses a time series y_n as a linear combination of past observations y_{n-i} and white noise v_{n-i} is called an *autoregressive moving average model* (ARMA model),

$$y_n = \sum_{i=1}^{m} a_i y_{n-i} + v_n - \sum_{i=1}^{\ell} b_i v_{n-i}. \qquad (6.1)$$

Here, m and a_i are called the *autoregressive order* and the *autoregressive coefficient* (AR coefficient), respectively. On the other hand, ℓ and b_i are called the *moving average order* and the *moving average coefficient* (MA coefficient), respectively. The AR order and the MA order (m, ℓ) taken together are called the ARMA order. Further, we assume that v_n is a white noise that follows a normal distribution with mean 0 and variance

σ^2, and is independent of the past time series y_{n-i}. That is, v_n satisfies:

$$\begin{array}{rcll} E[v_n] & = & 0, & \\ E[v_n^2] & = & \sigma^2, & \\ E[v_n v_m] & = & 0, & \text{for } n \neq m \\ E[v_n y_m] & = & 0, & \text{for } n > m. \end{array} \tag{6.2}$$

A time series y_n that follows an ARMA model is called an ARMA process. In practical terms, the most important model is an *AR model* (autoregressive model) of order m that expresses the time series as a linear combination of the past values y_{n-i} and the white noise v_n and is obtained by putting $\ell = 0$,

$$y_n = \sum_{i=1}^{m} a_i y_{n-i} + v_n. \tag{6.3}$$

On the other hand, the model obtained by putting $m = 0$,

$$y_n = v_n - \sum_{i=1}^{\ell} b_i v_{n-i}, \tag{6.4}$$

is called the *moving average model* (MA model) of order ℓ.

It should be noted that almost all analysis of stationary time series can be achieved by using AR models.

6.2   The Impulse Response Function

Using the *time shift operator* (or *lag operator*) B defined by $B y_n \equiv y_{n-1}$, the ARMA model can be expressed as

$$\left(1 - \sum_{i=1}^{m} a_i B^i \right) y_n = \left(1 - \sum_{i=1}^{\ell} b_i B^i \right) v_n. \tag{6.5}$$

Here, let the AR operator and the MA operator be defined respectively by

$$a(B) \equiv \left(1 - \sum_{i=1}^{m} a_i B^i \right), \quad b(B) \equiv \left(1 - \sum_{i=1}^{\ell} b_i B^i \right),$$

then the ARMA model can be concisely expressed as

$$a(B) y_n = b(B) v_n. \tag{6.6}$$

Dividing both sides of (6.6) by $a(B)$, the ARMA model can be expressed as $y_n = a(B)^{-1}b(B)v_n$. Therefore, if we define $g(B)$ as a formal infinite series

$$g(B) \equiv a(B)^{-1}b(B) = \sum_{i=0}^{\infty} g_i B^i, \tag{6.7}$$

the time series y_n that follows the ARMA model can be expressed by a moving average model with infinite order

$$y_n = g(B)v_n = \sum_{i=0}^{\infty} g_i v_{n-i}, \tag{6.8}$$

i.e. a linear combination of present and past realizations of white noise v_n.

The coefficients g_i; $i = 0, 1, \cdots$, correspond to the influence of the noise at time $n = 0$ to the time series at time i, and are called the *impulse response function* of the ARMA model. Here the impulse response g_i is obtained by the following recursive formula:

$$\begin{aligned} g_0 &= 1 \\ g_i &= \sum_{j=1}^{i} a_j g_{i-j} - b_i, \qquad i = 1, 2, \cdots, \end{aligned} \tag{6.9}$$

where $a_j = 0$ for $j > m$ and $b_j = 0$ for $j > \ell$.

Example   Consider the following four models.

(a) The first order AR model: $y_n = 0.9y_{n-1} + v_n$
(b) The second order AR model: $y_n = 0.9\sqrt{3}y_{n-1} - 0.81y_{n-2} + v_n$
(c) The second order MA model: $y_n = v_n - 0.9\sqrt{2}v_{n-1} + 0.81v_{n-2}$
(d) The ARMA model with order (2,2):
$\qquad y_n = 0.9\sqrt{3}y_{n-1} - 0.81y_{n-2} + v_n - 0.9\sqrt{2}v_{n-1} + 0.81v_{n-2}$

The plots (a), (b) (c) and (d) of Figure 6.1 show the impulse response functions obtained from (6.9) for the four models. The impulse response function of the MA model is non-zero only for the initial ℓ points. On the other hand, if the model contains an AR part, the impulse response function has non-zero values even for $i > m$, although it gradually decays.

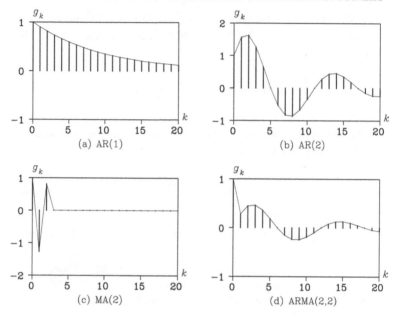

Figure 6.1: *Impulse response functions of four models.*

6.3 The Autocovariance Function

Taking the expectation after multiplying by y_{n-k} on both sides of (6.1), yields

$$E[y_n y_{n-k}] = \sum_{i=1}^{m} a_i E[y_{n-i} y_{n-k}] + E[v_n y_{n-k}] - \sum_{i=1}^{\ell} b_i E[v_{n-i} y_{n-k}]. \quad (6.10)$$

Here, from the expression of the ARMA model using the impulse response given in (6.7), the covariance between the time series y_m and the white noise v_n is given by

$$E[v_n y_m] = \sum_{i=0}^{\infty} g_i E[v_n v_{m-i}] = \begin{cases} 0 & n > m \\ \sigma^2 g_{m-n} & n \le m \end{cases}. \quad (6.11)$$

Substituting this into (6.10), we obtain the following equation with respect to the autocovariance function $C_k \equiv E[y_n y_{n-k}]$

$$C_0 = \sum_{i=1}^{m} a_i C_i + \sigma^2 \left(1 - \sum_{i=1}^{\ell} b_i g_i \right) \quad (6.12)$$

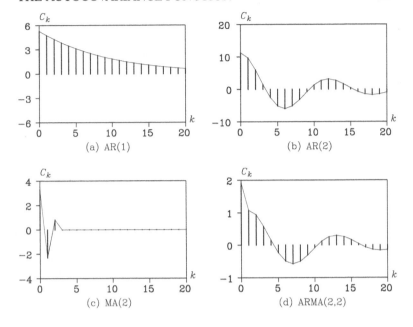

Figure 6.2: *Autocovariance functions of the four models.*

$$C_k = \sum_{i=1}^{m} a_i C_{k-i} - \sigma^2 \sum_{i=1}^{\ell} b_i g_{i-k}, \qquad k = 1, 2, \cdots.$$

Therefore, if the orders m and ℓ, the autoregressive and moving average coefficients a_i and b_i, and the innovation variance σ^2 of the ARMA model are given, we can obtain the autocovariance function by first computing the impulse response function $g_1, \cdots, g_\ell$ by (6.9) and then obtaining the autocovariance function $C_0, C_1, \cdots$ by (6.13). In particular, the following equation for the AR model obtained by putting $\ell = 0$ is called the *Yule-Walker equation*

$$C_0 = \sum_{i=1}^{m} a_i C_i + \sigma^2$$

$$C_k = \sum_{i=1}^{m} a_i C_{k-i}. \qquad (6.13)$$

Note that, since for univariate time series, the autocovariance function satisfies $C_{-k} = C_k$, the backward model satifies the same equation.

Example.   Figure 6.2 shows the autocovariance functions of the four models (a), (b), (c) and (d) shown in Figure 6.1. The autocovariance functions of (b) and (d) show a damped oscillation. On the other hand, for the MA model shown in (c), the autocovariance function becomes $C_k = 0$ for $k > 2$.

6.4   The Relation Between AR Coefficients and PARCOR

Hereafter the AR coefficient a_i of the AR model with order m is denoted as a_i^m. In particlar, a_m^m is called the m-th PARCOR (partial autocorrelation coefficient). As shown in Appendix B, the following relation holds between a_i^m and a_i^{m-1}

$$a_i^m = a_i^{m-1} - a_m^m a_{m-i}^{m-1}, \qquad i = 1, \cdots, m-1. \qquad (6.14)$$

If the M PARCOR's, $a_1^1, \cdots, a_M^M$, are given, repeated application of equation (6.14) yields the entire set of coefficients of the AR models with orders 2 through M. On the other hand, it can be seen from (6.14) that, if the coefficients $a_1^m, \cdots, a_m^m$ of the AR model of the highest order are given, by solving the equations

$$a_j^m = a_j^{m-1} - a_m^m \left(a_{m-j}^m + a_m^m a_j^{m-1} \right) \qquad (6.15)$$

for $j = i$, the coefficients of the AR model with order $m-1$ are obtained by

$$a_i^{m-1} = \frac{a_i^m + a_m^m a_{m-i}^m}{1 - (a_m^m)^2}. \qquad (6.16)$$

The PARCOR's, $a_1^1, \cdots, a_m^m$, can be obtained by repeating the above computation. This reveals that estimation of the coefficients $a_1^m, \cdots, a_m^m$ of the AR model of order m is equivalent to estimation of the PARCOR's up to the order m, i.e., $a_1^1, \cdots, a_m^m$.

Example   Figure 6.3 shows the PARCOR's of the four models (a), (b), (c) and (d). Contrary to the autocovariance function, if $i > m$ $a_i^i = 0$ for the AR model with order m. On the other hand, in the cases of the MA model and the ARMA model the PARCOR's gradually decay.

6.5   The Power Spectrum of the ARMA Process

If an ARMA model of a time series is given, the power spectrum of the time series can be obtained easily. Actually, the *power spectrum* of the

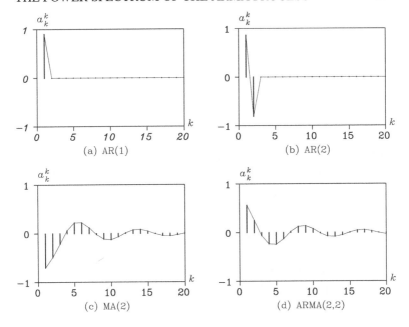

Figure 6.3: *PARCOR's of the four models.*

ARMA process (6.1) can be obtained from (6.8) as

$$
\begin{aligned}
p(f) &= \sum_{k=-\infty}^{\infty} C_k e^{-2\pi i k f} \qquad\qquad (6.17)\\
&= \sum_{k=-\infty}^{\infty} \mathrm{E}[y_n y_{n-k}] e^{-2\pi i k f}\\
&= \sum_{k=-\infty}^{\infty} \mathrm{E}\left[\left(\sum_{j=0}^{\infty} g_j v_{n-j}\right)\left(\sum_{p=0}^{\infty} g_p v_{n-k-p}\right)\right] e^{-2\pi i k f}\\
&= \sum_{k=-\infty}^{\infty}\sum_{j=0}^{\infty}\sum_{p=0}^{\infty} g_j g_p \mathrm{E}\left[v_{n-j} v_{n-k-p}\right] e^{-2\pi i k f}.
\end{aligned}
$$

Here using $g_p = 0$ for $p < 0$ and (6.2), the power spectrum is expressed as

$$
p(f) = \sigma^2 \sum_{k=-\infty}^{\infty}\sum_{j=0}^{\infty} g_j g_{j-k} e^{-2\pi i k f}
$$

$$= \sigma^2 \sum_{j=0}^{\infty} \sum_{k=-\infty}^{j} g_j e^{-2\pi i j f} g_{j-k} e^{-2\pi i (k-j) f}$$

$$= \sigma^2 \sum_{j=0}^{\infty} \sum_{p=0}^{\infty} g_j e^{-2\pi i j f} g_p e^{2\pi i p f}$$

$$= \sigma^2 \left| \sum_{j=0}^{\infty} g_j e^{-2\pi i j f} \right|^2 , \tag{6.18}$$

where $\sum_{j=0}^{\infty} g_j e^{-2\pi i j f}$ is the Fourier transform of the impulse response function, and is called the *frequency response function*. On the other hand, putting $B = e^{-2\pi i f}$ in (6.7), it can be expressed as

$$\sum_{j=0}^{\infty} g_j e^{-2\pi i j f} = \left(1 - \sum_{j=1}^{m} a_j e^{-2\pi i j f} \right)^{-1} \left(1 - \sum_{j=1}^{\ell} b_j e^{-2\pi i j f} \right). \tag{6.19}$$

Therefore, substituting the above frequency response function into (6.18), the power spectrum of the ARMA model is given by

$$p(f) = \sigma^2 \frac{\left| 1 - \sum_{j=1}^{\ell} b_j e^{-2\pi i j f} \right|^2}{\left| 1 - \sum_{j=1}^{m} a_j e^{-2\pi i j f} \right|^2}. \tag{6.20}$$

Example. Figure 6.4 shows the logarithm of the power spectra of the four models. The power spectrum of the AR model with order one does not have any peak or trough. A peak is seen in the plot (b) of the spectrum of the second order AR model and one trough is seen in the plot (c) of the second order MA model. On the other hand, the spectrum of the ARMA model of order (2,2) shown in the plot (d) has both one peak and one trough.

These examples indicate that there must be close relations between the AR and MA orders and the number of peaks and troughs in the spectra. Actually, the logarithm of the spectrum, $\log p(f)$ shown in Figure 6.4 is expressible as

$$\log p(f) = \log \sigma^2 - 2 \log \left| 1 - \sum_{j=1}^{m} a_j e^{-2\pi i j f} \right| + 2 \log \left| 1 - \sum_{j=1}^{\ell} b_j e^{-2\pi i j f} \right|.$$

$$\tag{6.21}$$

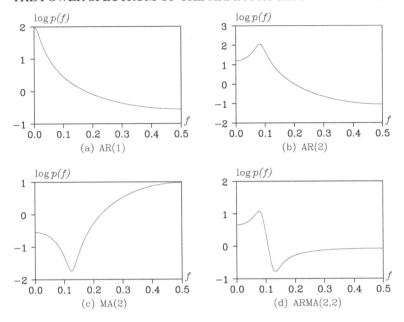

Figure 6.4: *Logarithm of the power spectra of the four models.*

Therefore, the peak and the trough of the spectrum appear at the local minimum of $|1 - \sum_{j=1}^{m} a_j e^{-2\pi i j f}|$ and $|1 - \sum_{j=1}^{l} b_j e^{-2\pi i j f}|$, respectively. The number of peaks and troughs respectively correspond to the number of roots of the AR operator and the MA operator as will be explained in the next section. To express k peaks or k troughs, the AR order or the MA order must be higher than or equal to $2k$, respectively. Moreover, the locations and the heights of the peaks or the troughs are related to the angles and the absolute values of the complex roots of the characteristic equation.

In particular, when the angles of the complex roots of the AR operator coincide with those of the MA operator, a line-like spectral peak or trough appears. For example, if the AR and MA coefficients of the ARMA (2,2) model are given by

$$m = 2, \quad a_1 = 0.99\sqrt{2}, \quad a_2 = -0.99^2$$
$$\ell = 2, \quad b_1 = 0.95\sqrt{2}, \quad b_2 = -0.95^2,$$

both characteristic equations have roots at $f = 0.125$ (= 45 degrees), and $\log p(f)$ has a line-like spectral peak as shown in Figure 6.5.

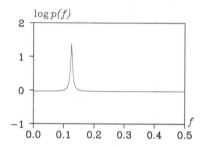

Figure 6.5: *Line-like spectrum of ARMA(2.2) model.*

6.6 The Characteristic Equation

The characteristics of an ARMA model are determined by the roots of the following two polynomial equations:

$$a(B) = 1 - \sum_{j=1}^{m} a_j B^j = 0 \qquad (6.22)$$

$$b(B) = 1 - \sum_{j=1}^{\ell} b_j B^j = 0. \qquad (6.23)$$

Equations (6.22) and (6.23) are called the *characteristic equation* associated with the AR operator, and the MA operator, respectively. The roots of these equations are called the *characteristic roots*. If the roots of the characteristic equation $a(B) = 0$ of the AR operator all lie outside the unit circle, the influence of noise turbulence at a certain time decays as time progresses, and the ARMA model becomes *stationary*.

On the other hand, if all roots of the characteristic equation $b(B) = 0$ of the MA operator lie outside the unit circle, the coefficient of h_i of $b(B)^{-1} = \sum_{i=0}^{\infty} h_i B^i$ converges and the ARMA model can be expressed by an AR model of infinite order as

$$y_n = - \sum_{i=1}^{\infty} h_i y_{n-i} + v_n. \qquad (6.24)$$

In this case, the time series is called *invertible*.

As mentioned in the previous section, the positions of the roots of the two characteristic polynomials have a close relation to the shape of the spectrum. The peak of the spectrum (or trough) appears at $f = \theta/2\pi$, if the complex root of AR (or MA) operator is expressed in the form

$$z = \alpha + i\beta = re^{i\theta}. \tag{6.25}$$

Further, the closer the root r approaches to 1, the sharper the peak or the trough of the spectrum become. Figure 6.6 shows the positions of the characteristic roots of the four models that have been used for the examples in this chapter. The symbols $*$ and $+$ denote the roots of the AR operator and the roots of the MA operator, respectively. For convenience in illustration, the position of $z^{-1} = r^{-1}e^{-i\theta}$ is displayed in Figure 6.6 instead of z.

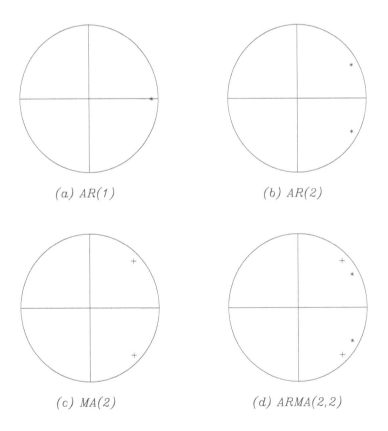

(a) AR(1)   (b) AR(2)

(c) MA(2)   (d) ARMA(2,2)

Figure 6.6 Characteristic roots. (a) AR model of order 1, (b) AR model of order 2, (c) MA model with order 2 and (d) ARMA model with order (2,2).

The function armachar of the package TSSS computes and draw graphs of the impulse response function, the autocovariance function, the PARCORs, power spectrum and the characteristic roots of the AR operator and the MA operator of an ARMA model. The following arguments are required

arcoef:	vector of AR coefficients.
macoef:	vector of MA coefficients.
v:	innovation variance.
lag:	maximum lag of autocovariance, etc. Default is 50.
nf:	number of frequencies in evaluating spectrum. Default is 200.

and the outputs from this function are:

impuls:	impulse response function.
acov:	autocovariance function.
parcor:	partial autocorrelation function.
spec:	power spectrum.
croot.ar:	characteristic roots of AR operator.
croot.ma:	characteristic roots of MA operator.

```
> # AR model : y(n) = a(1)*y(n-1) + a(2)*y(n-2) + v(n)
> a <- c(0.9 * sqrt(3), -0.81)
> armachar(arcoef = a, v = 1.0, lag = 20)

> # MA model : y(n) = v(n) - b(1)*v(n-1) - b(2)*v(n-2)
> b <- c(0.9 * sqrt(2), -0.81)
> armachar(macoef = b, v = 1.0, lag = 20)

> # ARMA model : y(n) = a(1)*y(n-1) + a(2)*y(n-2)
> # + v(n) - b(1)*v(n-1) - b(2)*v(n-2)
> armachar(arcoef = a, macoef = b, v = 1.0, lag = 20)

> # ARMA model with line spectrum
> a <- c(0.99 * sqrt(2),-0.9801)
> b <- c(0.95 * sqrt(2),-0.9025)
> armachar(arcoef = a, macoef = b, v = 1.0, lag = 20)
```

Figure 6.7 shows the output of the function armachar for the ARMA(2,2) model. Impulse response function, autocovariance function, PARCOR's, power spectrum in log-scale and characteristic roots are shown.

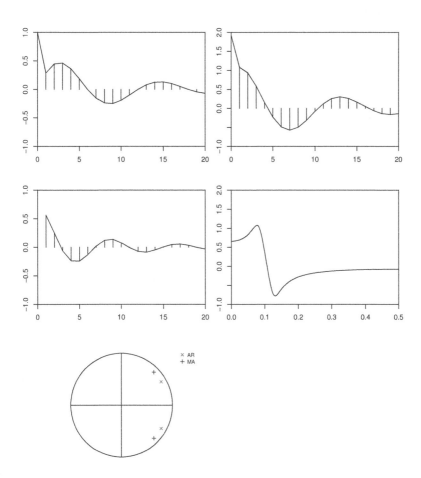

Figure 6.7 *armachar for ARMA(2,2). Impulse response function (top left), auto-covariance function (top right), PARCOR's (middle left), power spectrum in log-scale (middle right) and characteristic roots (lower).*

6.7   The Multivariate AR Model

Similar to the case of univariate time series, for *multivariate time series*, $y_n = (y_n(1), \cdots, y_n(\ell))^T$, the model that expresses a present value of the time series as a linear combination of past values $y_{n-1}, \cdots, y_{n-M}$ and white noise is called a *multivariate autoregressive model* (MAR model)

$$y_n = \sum_{m=1}^{M} A_m y_{n-m} + v_n, \tag{6.26}$$

where A_m is the autoregressive coefficient matrix whose (i, j)-th element is given by $a_m(i, j)$, and v_n is an ℓ-dimensional white noise that satisfies

$$E[v_n] = \begin{bmatrix} 0 \\ \vdots \\ 0 \end{bmatrix}, \qquad E[v_n v_n^T] = \begin{bmatrix} \sigma_{11} & \cdots & \sigma_{1\ell} \\ \vdots & \ddots & \vdots \\ \sigma_{\ell 1} & \cdots & \sigma_{\ell\ell} \end{bmatrix} = W$$

$$\begin{aligned} E[v_n v_m^T] &= O, \qquad \text{for } n \neq m \\ E[v_n y_m^T] &= O, \qquad \text{for } n > m. \end{aligned} \tag{6.27}$$

Here, O denotes the $\ell \times \ell$ matrix with 0 elements, and W is an $\ell \times \ell$ symmetric matrix satisfying $\sigma_{ij} = \sigma_{ji}$. The cross-covariance function of $y_n(i)$ and $y_n(j)$ is defined by $C_k(i, j) = E\left[y_n(i)y_{n-k}(j)\right]$. Then, the $\ell \times \ell$ matrix $C_k = E[y_n y_{n-k}^T]$, the (i, j)-th component of which is $C_k(i, j)$, is called the *cross-covariance function*. Similar to the case of the univariate time series, for the multivariate AR model, C_k satisfies the Yule-Walker equation

$$C_0 = \sum_{j=1}^{M} A_j C_{-j} + W \tag{6.28}$$

$$C_k = \sum_{j=1}^{M} A_j C_{k-j}, \qquad (k = 1, 2, \cdots). \tag{6.29}$$

As noted in Chapter 2, the cross-covariance function is not symmetric. Therefore, for multivariate time series, the Yule-Walker equations for the the backward AR model and the forward AR model are different.

The Fourier transform of the cross-covariance function $C_k(s, j)$

$$\begin{aligned} p_{sj}(f) &= \sum_{k=-\infty}^{\infty} C_k(s, j) e^{-2\pi i k f} \tag{6.30} \\ &= \sum_{k=-\infty}^{\infty} C_k(s, j) \cos(2\pi k f) - i \sum_{k=-\infty}^{\infty} C_k(s, j) \sin(2\pi k f) \end{aligned}$$

is called the *cross-spectral density function*. Since the cross-covariance function is not an even function, the cross-spectrum given by (6.31) has an imaginary part and is a complex number. The $\ell \times \ell$ cross-spectrum matrix $P(f)$ is defined by

$$P(f) = \begin{bmatrix} p_{11}(f) & \cdots & p_{1\ell}(f) \\ \vdots & \ddots & \vdots \\ p_{\ell 1}(f) & \cdots & p_{\ell\ell}(f) \end{bmatrix}, \tag{6.31}$$

and the relations between $P(f)$ and the cross-covariance matrix C_k are given by

$$P(f) = \sum_{k=-\infty}^{\infty} C_k e^{-2\pi ikf} \tag{6.32}$$

$$C_k = \int_{-\frac{1}{2}}^{\frac{1}{2}} P(f) e^{2\pi ikf} df. \tag{6.33}$$

For time series that follow the multivariate AR model, the cross-spectrum can be obtained by (Whittle (1963))

$$P(f) = A(f)^{-1} W (A(f)^{-1})^*, \tag{6.34}$$

where A^* denotes the complex conjugate of the matrix A, and $A(f)$ denotes the $\ell \times \ell$ matrix whose (j,k)-th component is defined by

$$A_{jk}(f) = \sum_{m=0}^{M} a_m(j,k) e^{-2\pi imf}. \tag{6.35}$$

Here, it is assumed that $a_0(j,j) = -1$ and $a_0(j,k) = 0$ for $j \neq k$. Given a frequency f, the cross-spectrum is a complex number and can be expressed as

$$p_{jk}(f) = \alpha_{jk}(f) e^{i\phi_{jk}(f)}, \tag{6.36}$$

where

$$\alpha_{jk}(f) = \sqrt{(\Re\{p_{jk}(f)\})^2 + (\Im\{p_{jk}(f)\})^2}$$

$$\phi_{jk}(f) = \arctan\left\{ \frac{\Im\{p_{jk}(f)\}}{\Re\{p_{jk}(f)\}} \right\}.$$

$\Re$ and $\Im$ denote the real and imaginary parts of the complex number, respectively. Then, $\alpha_{jk}(f)$ is called the *amplitude spectrum* and $\phi_{jk}(f)$

is the *phase spectrum*. Moreover,

$$coh_{jk}(f) = \frac{\alpha_{jk}(f)^2}{p_{jj}(f)p_{kk}(f)} \tag{6.37}$$

denotes the square of the correlation coefficient between frequency components of time series $y_n(j)$ and $y_n(k)$ at frequency f and is called the *coherency*.

For convenience, in the following $A(f)^{-1}$ will be denoted as $B(f) = (b_{jk}(f))$. If the components of the white noise v_n are mutually uncorrelated and the variance covariance matrix is the diagonal matrix $W = \text{diag}\{\sigma_1^2, \cdots, \sigma_\ell^2\}$, then the power spectrum of the i-th component of the time series can be expressed as

$$p_{ii}(f) = \sum_{j=1}^{\ell} b_{ij}(f)\sigma_j^2 b_{ij}(f)^* \equiv \sum_{j=1}^{\ell} |b_{ij}(f)|^2 \sigma_j^2. \tag{6.38}$$

This indicates that the power of the fluctuation of the component i at frequency f can be decomposed into the effects of ℓ noises, i.e. $|b_{ij}(f)|^2\sigma_j^2$. Therefore, if we define $r_{ij}(f)$ by

$$r_{ij}(f) = \frac{|b_{ij}(f)|^2\sigma_j^2}{p_{ii}(f)}; \tag{6.39}$$

it represents the ratio of the effect of $v_n(j)$ in the fluctuation of $y_n(i)$ at frequency f.

This quantity $r_{ij}(f)$ is called the *relative power contribution* which is a useful tool for the analysis of a feedback system (Akaike (1968), Akaike and Nakagawa (1989), Ohtsu et al. (2015)).

Example

The function `marspc` of the package TSSS computes the cross-spectra, the coherency and the power contribution of multivariate time series. The following arguments are required

    `arcoef`:   AR coefficient matrices.
    `v`:   innovation covariance matrix.
and the outputs from this function are:

    `spec`:   cross-spectra.
    `amp`:   amplitude spectra.
    `phase`:   phase spectra.
    `coh`:   simple coherency.
    `power`:   power contribution.
    `rpowor`:   relative power contribution.

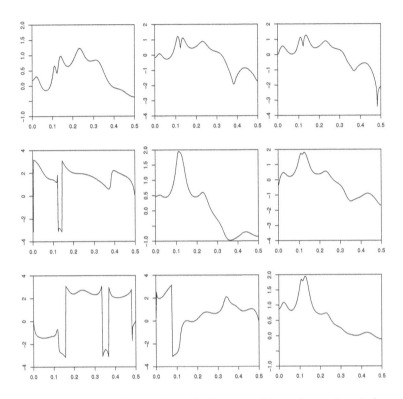

Figure 6.8 *Spectra (diagonal), amplitude spectra (above diagonal) and phase spectra (below diagonal) for the three variate ship data (yaw rate, pitch rate and rudder angle).*

```
> data(HAKUSAN)
> yy <- as.matrix(HAKUSAN[, c(1,2,4)])
> nc <- dim(yy)[1]
> n <- seq(1, nc, by = 2)
> y <- yy[n, ]
> # Fit MAR model
> z <- marfit(y, lag = 20)
> # Draw cross-spectra, coherency, power contribution
> marspc(z$arcoef, v = z$v)
```

Figure 6.8 shows the cross-spectra obtained by using a multivariate AR model for the three-variate time series composed of the yaw rate, the role rate and the rudder angle shown in (a) and (h) of Figure 1.1 ($N = 500$

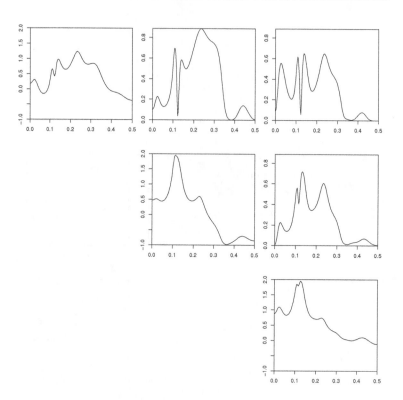

Figure 6.9 *Power spectra (3 in diagonal) and coherencies for three variate ship data (yaw rate, role rate and rudder angle).*

and $\triangle t = 2$ second) that were originally sampled every second. Three of nine plots on the diagonal in the figure show the logarithm of the power spectra of the yaw rate, the role rate and the rudder angle, respectively. As for the power spectra of the yaw rate, the maximum peak is seen in the vicinity of $f = 0.25$ (8 seconds cycle) and for the role rate and the rudder angle in the vicinity of $f = 0.125$ (16 seconds cycle). On the other hand, three plots above the diagonal show the absolute values of the amplitude spectra of the cross-spectra, that is, the logarithm of the amplitude spectra. Three plots below the diagonal show the phase spectra where some discontinuous jumps are seen. The reason for this is that the phase spectra are displayed within the range $[-\pi, \pi]$.

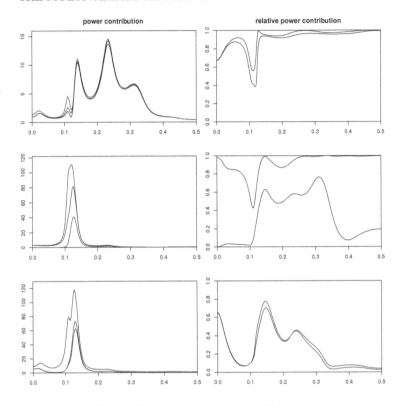

Figure 6.10: *Power contributions of ship's data.*

Figure 6.9 shows the power spectra and the coherencies. Three plots
on the diagonal show the power spectra similarly to Figure 6.8. Three
other plots above the diagonal show the coherencies. Rolling and rudder
angle have two significant peaks at the same frequencies. The peak of
the yaw rate has slightly smaller frequency than those of yaw rate and
rolling.

On the other hand, Figure 6.10 shows the power contributions. In this
figure, the plots in the left column show the absolute power contribution
and the plots in the right column show the relative power contribution.
All plots in both columns of Figure 6.10 show the contribution of the yaw
rate, the role rate and the rudder angle from the bottom to the top in each
plot, respectively. The influence of the rudder angle is clearly visible for
$f < 0.13$. However, the influence of the rudder angle is barely noticeable

in the vicinity of $f = 0.14$ and 0.23, where the dominant power of the yaw rate is found. This is probably explained by noting that this data set has been observed under the control of a conventional PID autopilot system that is designed to suppress the power of variation in the frequency area $f < 0.13$.

Two plots in the second row show the power contribution of the role rate. The contribution of the rudder angle is almost 50 percent in the vicinity of $f = 0.1$ where the power spectrum is significant. This is thought to be a side effect of the steering to suppress the variation of the yaw rate. On the other hand, for $f > 0.14$, a strong influence of the yaw rate is seen, although the actual power is very small.

Two figures in the third row show the power contribution to the rudder angle. It can be seen that the influence of the yaw rate is very strong in the vicinity of $f = 0.12$ where the power spectrum is large. Moreover, it can also be seen that in the range of $f < 0.08$, the contribution of the yaw rate becomes greater as the frequency decreases.

Problems

1. (1) Show the stationarity condition for AR(1).
   (2) Show the stationarity condition for AR(2).

2. For an AR(1), $y_n = ay_{n-1} + v_n, \quad v_n \sim N(0, \sigma^2)$:
   (1) Obtain the one-step-ahead prediction error variance.
   (2) Obtain the two-step-ahead prediction error variance.
   (3) Obtain the k-step-ahead prediction error variance.

3. Assuming that the time series follows the models shown below and that v_n follows a white noise with mean 0 and variance σ^2, obtain the autocovariance function C_k.
   (1) AR model of order 1: $y_n = -0.9y_{n-1} + v_n$.
   (2) AR model of order 2: $y_n = 1.2y_{n-1} - 0.6y_{n-2} + v_n$.
   (3) MA model of order 1: $y_n = v_n - bv_{n-1}$
   (4) ARMA model of order (1,1): $y_n = ay_{n-1} + v_n - bv_{n-1}$.

4. Assume that a time series follows an AR model of order 1, $y_n = ay_{n-1} + v_n, v_n \sim N(0, 1)$.
   (1) When the noise term v_n is not a white noise but follows an autoregressive process of order 1, $v_n = bv_{n-1} + w_n$, show that y_n follows an AR model of order 2.

(2) Obtain the autocovariance function C_k, $k = 0,1,2,3$ of the contaminated series x_n defined by $x_n = y_n + w_n$, $w_n \sim N(0,0.1)$.

5. (1) Using the result of Problem 3 for Chapter 3 and the definition of the power spectrum, show that the power spectrum of MA model with order 1 can be expressed as $p(f) = |1 - be^{-2\pi i f}|^2$, where the right-hand side can be expressed as $1 + b^2 - 2b\cos(2\pi f)$.

(2) Using the fact that if $\sigma^2 = 1$, the spectrum of an AR model of order 1, $y_n = ay_{n-1} + v_n$, can be expressed as $p(f) = (1 - 2a\cos(2\pi f) + a^2)^{-1}$, show that the maximum and the minimum of the spectrum occurs at $f = 0$ or $f = 0.5$. Also, consider where the spectrum $p(f)$ attains its maximum.

6. For an AR model of order 2, $y_n = a_1 y_{n-1} + a_2 y_{n-2} + v_n$, $v_n \sim N(0,\sigma^2)$;

(1) Show the formula to obtain the autocovariance function $C_0, C_1, \ldots$.

(2) Show the expression to obtain the power spectrum $p(f)$, $0 \le f \le 1/2$.

(3) Obtain C_0, C_1, C_2 when $a_1 = 0.8$, $a_2 = -0.6$ and $\sigma^2 = 1$.

(4) For the same case, obtain the expression for the power spectrum $p(f)$. Investigate for which frequency f, $p(f)$ attains its maximum.

7. Assume that a time series y_n follows an MA model of order 1; $y_n = v_n - bv_{n-1}$, $v_n \sim N(0,1)$.

(1) Obtain the autocovariance function C_k, $k = 0,1,2,3$.

(2) Express the time series by an AR model.

8. (1) Obtain the variance of the k-step-ahead prediction error $\varepsilon_{n+k|n}$ for an MA model of order 1; $y_n = v_n - bv_{n-1}$, $v_n \sim N(0,1)$.

(2) Express an AR model of order 1, $y_n = ay_{n-1} + v_n$, using an MA model of infinite order and obtain the variance of the k-step-ahead prediction error variance.

(3) Using the formal expansion of the random walk model $y_n = y_{n-1} + v_n$, obtain the MA model of infinite order. Using that expression, obtain the k-step-ahead prediction error variance of the random walk model.

Chapter 7

Estimation of an AR Model

Among the stationary time series models discussed in the preceding chapter, very efficient estimation methods can be derived for AR models. This chapter presents methods for estimating the parameters of the AR model by the Yule-Walker method, the least squares method and the PARCOR methods. The method of determining the order of the AR model using the AIC is also shown. In addition, the Yule-Walker method and the least squares method for the estimation and identification of the multivariate AR model are shown.

7.1 Fitting an AR Model

Given a time series $y_1, \cdots, y_N$, we consider the problem of fitting an *autoregressive model* (AR model)

$$y_n = \sum_{i=1}^{m} a_i y_{n-i} + v_n, \tag{7.1}$$

where m is the *order* of the autoregression, a_i is the *autoregressive coefficient* and v_n is the white noise that follows a normal distribution with mean 0 and variance σ^2 (Akaike (1969), Box and Jenkins (1970), Akaike and Nakagawa (1989), Brockwell and Davis (1991)). σ^2 is often called the *innovation variance*.

To identify an AR model, it is necessary to determine the order m and estimate the autoregressive coefficients $a_1, \cdots, a_m$ and the variance σ^2 from the data. In the following, these parameters will be denoted by $\theta = (a_1, \cdots, a_m, \sigma^2)^T$.

Under the assumption that the order m is given, consider the estimation of the parameter θ by the maximum likelihood method. The joint distribution of time series $y = (y_1, \cdots, y_N)^T$ following the AR model becomes a multivariate normal distribution. When the model (7.1) is given, the mean vector of the autoregressive process y is $(0, \ldots, 0)^T$ and the variance covariance matrix is given by

$$\Sigma = \begin{bmatrix} C_0 & C_1 & \cdots & C_{N-1} \\ C_1 & C_0 & \cdots & C_{N-2} \\ \vdots & \vdots & \ddots & \vdots \\ C_{N-1} & C_{N-2} & \cdots & C_0 \end{bmatrix}, \tag{7.2}$$

where the autocovariance function C_k is defined by (6.12). Therefore, the likelihood of the AR model is obtained by

$$\begin{aligned} L(\theta) &= p(y_1, \cdots, y_N | \theta) \\ &= (2\pi)^{-\frac{N}{2}} |\Sigma|^{-\frac{1}{2}} \exp\left\{ -\frac{1}{2} y^T \Sigma^{-1} y \right\}. \tag{7.3} \end{aligned}$$

However, when the number of data, N, is large, the computation of the likelihood by this method becomes difficult because it involves the inversion and computation of the determinant of the $N \times N$ matrix Σ. Further, to obtain the maximum likelihood estimate of θ that maximizes (7.3), it is necessary to apply a numerical optimization method, since the likelihood is a complicated function of the parameter θ. In general, however, the likelihood of a time series model can be efficiently calculated by expressing it as a product of conditional distributions

$$\begin{aligned} L(\theta) &= p(y_1, \cdots, y_N | \theta) \\ &= p(y_1, \cdots, y_{N-1} | \theta) p(y_N | y_1, \cdots, y_{N-1}, \theta) \\ &\quad \vdots \\ &= \prod_{n=1}^{N} p(y_n | y_1, \cdots, y_{n-1}, \theta). \tag{7.4} \end{aligned}$$

Using a Kalman filter, each term in the right-hand side of (7.4) can be efficiently and exactly evaluated, which makes it possible to compute the exact likelihood of the ARMA model and other time series models. Such a method will be treated in Chapter 9. When the *maximum likelihood estimate* $\hat{\theta}$ of the AR model has been obtained, the AIC for the model is defined by

$$\begin{aligned} \text{AIC} &= -2\,(\text{maximum log-likelihood}) + 2\,(\text{number of parameters}) \\ &= -2 \log L(\hat{\theta}) + 2(m+1). \tag{7.5} \end{aligned}$$

To select the AR order m by the minimum AIC method, we calculate the AIC's of the AR models with orders up to M, i.e., $\text{AIC}_0, \cdots, \text{AIC}_M$, and select the order that results in the minimum AIC value (Akaike (1973,1974), Sakamoto et al. (1986), Konishi and Kitagawa (2008)).

7.2 Yule-Walker Method and Levinson's Algorithm

As shown in Chapter 6, the autocovariance function of the AR model (7.1) of order m satisfies the *Yule-Walker equation* (Akaike (1969), Box and Jenkins (1970))

$$C_0 = \sum_{i=1}^{m} a_i C_i + \sigma^2 \tag{7.6}$$

$$C_j = \sum_{i=1}^{m} a_i C_{j-i}. \tag{7.7}$$

On the other hand, once the time series has been obtained, by computing the sample autocovariance functions $\hat{C}_k$ and substituting them into (7.7), we obtain a system of linear equations for the unknown autoregressive coefficients $a_1, \cdots, a_m$,

$$
\begin{bmatrix}
\hat{C}_0 & \hat{C}_1 & \cdots & \hat{C}_{m-1} \\
\hat{C}_1 & \hat{C}_0 & \cdots & \hat{C}_{m-2} \\
\vdots & \vdots & \ddots & \vdots \\
\hat{C}_{m-1} & \hat{C}_{m-2} & \cdots & \hat{C}_0
\end{bmatrix}
\begin{bmatrix}
a_1 \\
a_2 \\
\vdots \\
a_m
\end{bmatrix}
=
\begin{bmatrix}
\hat{C}_1 \\
\hat{C}_2 \\
\vdots \\
\hat{C}_m
\end{bmatrix}. \tag{7.8}
$$

The estimates $\hat{a}_i$ of the AR coefficients are obtained by solving this equation. Then from (7.6), an estimate of the variance σ^2 is obtained by

$$\hat{\sigma}^2 = \hat{C}_0 - \sum_{i=1}^{m} \hat{a}_i \hat{C}_i. \tag{7.9}$$

The estimates $\hat{a}_1, \cdots, \hat{a}_m$, and $\hat{\sigma}^2$ obtained by this method are called the *Yule-Walker estimates*. Since the variance of the prediction errors of the AR model with coefficients a_i is given by

$$
\begin{aligned}
E[v_n^2] &= E\left[\left(y_n - \sum_{i=1}^{m} a_i y_{n-i}\right)^2\right] \\
&= C_0 - 2\sum_{i=1}^{m} a_i C_i + \sum_{i=1}^{m}\sum_{j=1}^{m} a_i a_j C_{i-j}, \tag{7.10}
\end{aligned}
$$

we obtain equation (7.7) from

$$\frac{\partial E[v_n^2]}{\partial a_j} = -2C_j + 2\sum_{i=1}^{m} a_i C_{j-i} = 0. \tag{7.11}$$

Therefore, we can consider that the Yule-Walker estimates obtained by solving (7.8) after substituting $\hat{C}_j$ for C_j in (7.11) approximately minimize the variance of prediction errors.

To obtain the Yule-Walker estimates for an AR model of order m, it is necessary to solve a system of linear equations with m unknowns. In addition, to select the order of the AR model by the minimum AIC method, we need to evaluate the AIC values of the models with orders up to the maximum order, M. Namely, we have to estimate the coefficients by solving systems of linear equations with 1 unknown, ... , M unknowns.

However, by the *Levinson's algorithm*, these solutions can be obtained quite efficiently. Hereafter, the AR coefficients and the innovation variance of the AR model of order m are denoted as a_j^m and σ_m^2, respectively. Then Levinson's algorithm is defined as follows:

1. Set $\hat{\sigma}_0^2 = \hat{C}_0$ and $\text{AIC}_0 = N(\log 2\pi\hat{\sigma}_0^2 + 1) + 2$
2. For $m = 1, \cdots, M$, repeat the following steps

    (a) $\hat{a}_m^m = \left(\hat{C}_m - \sum_{j=1}^{m-1} \hat{a}_j^{m-1}\hat{C}_{m-j}\right)\left(\hat{\sigma}_{m-1}^2\right)^{-1}$,

    (b) $\hat{a}_i^m = \hat{a}_i^{m-1} - \hat{a}_m^m\hat{a}_{m-i}^{m-1}$ for $i = 1,\ldots,m-1$,

    (c) $\hat{\sigma}_m^2 = \hat{\sigma}_{m-1}^2\{1 - (\hat{a}_m^m)^2\}$,

    (d) $\text{AIC}_m = N(\log 2\pi\hat{\sigma}_m^2 + 1) + 2(m+1)$.

In Levinson's algorithm, the PARCOR $\hat{a}_m^m$ introduced in Chapter 6 plays an important role. This algorithm will be explained in detail in Appendix B.

7.3 Estimation of an AR Model by the Least Squares Method

In this section, the least squares method explained in Chapter 5 will be applied to the estimation of the AR model. Putting $\theta = (a_1, \cdots, a_m, \sigma^2)^T$, from (7.4), the log-likelihood of the AR model becomes

$$\ell(\theta) = \sum_{n=1}^{N} \log p(y_n|y_1, \cdots, y_{n-1}). \qquad (7.12)$$

Here, for the AR model of order m, since the distribution of y_n is specified by the values of $y_{n-1}, \cdots, y_{n-m}$ for $n > m$, each term in (7.12) is given by

$$p(y_n|y_1, \cdots, y_{n-1}) = p(y_n|y_{n-m}, \cdots, y_{n-1})$$

$$= -\frac{1}{\sqrt{2\pi\sigma^2}} \exp\left\{ -\frac{1}{2\sigma^2} \left(y_n - \sum_{i=1}^{m} a_i y_{n-i} \right)^2 \right\}$$

$$\log p(y_n | y_1, \cdots, y_{n-1}) = -\frac{1}{2}\log 2\pi\sigma^2 - \frac{1}{2\sigma^2}\left(y_n - \sum_{i=1}^{m} a_i y_{n-i} \right)^2.$$
$$(7.13)$$

Therefore, by ignoring the initial M $(M \geq m)$ terms of (7.12), an approximate log-likelihood of the AR model is obtained as

$$\ell(\theta) = -\frac{N-M}{2}\log 2\pi\sigma^2 - \frac{1}{2\sigma^2} \sum_{n=M+1}^{N} \left(y_n - \sum_{i=1}^{m} a_i y_{n-i} \right)^2 \quad (7.14)$$

(Kitagawa and Akaike (1978), Sakamoto et al. (1986), Kitagawa and Gersch (1996)).

Similar to the case of the regression model, for arbitrarily given autoregressive coefficients $a_1, \cdots, a_m$, an approximate maximum likelihood estimate of the variance σ^2 maximizing (7.14) satisfies

$$\frac{\partial \ell(\theta)}{\partial \sigma^2} = -\frac{N-M}{2\sigma^2} + \frac{1}{2(\sigma^2)^2} \sum_{n=M+1}^{N} \left(y_n - \sum_{i=1}^{m} a_i y_{n-i} \right)^2 = 0, \quad (7.15)$$

and is obtained as

$$\hat{\sigma}^2 = \frac{1}{N-M} \sum_{n=M+1}^{N} \left(y_n - \sum_{i=1}^{m} a_i y_{n-i} \right)^2. \quad (7.16)$$

Substituting this into (7.14), the approximate log-likelihood becomes a function of the autoregressive coefficients $a_1, \cdots, a_m$

$$\ell(a_1, \cdots, a_m) = -\frac{N-M}{2}\log 2\pi\hat{\sigma}^2 - \frac{N-M}{2}. \quad (7.17)$$

Here, since the logarithm is a monotone increasing function, maximization of the approximate log-likelihood (7.17) can be achieved by minimizing the variance $\hat{\sigma}^2$. This means that the approximate maximum likelihood estimates of the AR model can be obtained by the least squares method. To obtain the least squares estimates of the AR models with orders up to M by the Householder transformation discussed in Chapter 5, define the matrix Z and the vector y by

$$Z = \begin{bmatrix} y_M & y_{M-1} & \cdots & y_1 \\ y_{M+1} & y_M & \cdots & y_2 \\ \vdots & \vdots & \ddots & \vdots \\ y_{N-1} & y_{N-2} & \cdots & y_{N-M} \end{bmatrix}, \qquad y = \begin{bmatrix} y_{M+1} \\ y_{M+2} \\ \vdots \\ y_N \end{bmatrix}. \quad (7.18)$$

For actual computation, construct the $(N - M) \times (M + 1)$ matrix

$$
X = [Z \mid y] =
\begin{bmatrix}
y_M & \cdots & y_1 & y_{M+1} \\
y_{M+1} & \cdots & y_2 & y_{M+2} \\
\vdots & \ddots & \vdots & \vdots \\
y_{N-1} & \cdots & y_{N-M} & y_N
\end{bmatrix},
\qquad (7.19)
$$

and transform it to an upper triangular matrix

$$
HX = \begin{bmatrix} S \\ O \end{bmatrix} =
\begin{bmatrix}
s_{11} & \cdots & s_{1M} & s_{1,M+1} \\
& \ddots & \vdots & \vdots \\
& & s_{MM} & s_{M,M+1} \\
& & & s_{M+1,M+1} \\
& & O &
\end{bmatrix},
\qquad (7.20)
$$

by Householder transformation.

Then, for $0 \le j \le M$, the innovation variance and the AIC of the AR model of order j are obtained by

$$
\begin{aligned}
\hat{\sigma}_j^2 &= \frac{1}{N-M} \sum_{i=j+1}^{M+1} s_{i,M+1}^2 \\
\text{AIC}_j &= (N-M)(\log 2\pi\hat{\sigma}_j^2 + 1) + 2(j+1). \qquad (7.21)
\end{aligned}
$$

Moreover, the least squares estimates of the autoregressive coefficients, which are the solutions of the linear equation

$$
\begin{bmatrix}
s_{11} & \cdots & s_{1j} \\
& \ddots & \vdots \\
& & s_{jj}
\end{bmatrix}
\begin{bmatrix}
a_1 \\
\vdots \\
a_j
\end{bmatrix} =
\begin{bmatrix}
s_{1,M+1} \\
\vdots \\
s_{j,M+1}
\end{bmatrix},
\qquad (7.22)
$$

can be easily obtained by backward substitution as follows:

$$
\begin{aligned}
\hat{a}_j &= \frac{s_{j,M+1}}{s_{jj}} \\
\hat{a}_i &= \frac{s_{i,M+1} - s_{i,i+1}\hat{a}_{i+1} - \cdots - s_{i,j}\hat{a}_j}{s_{jj}}, \qquad i = j-1,\cdots,1.
\end{aligned}
$$

7.4 Estimation of an AR Model by the PARCOR Method

Assuming that the autocovariance functions $C_0, C_1, \cdots$ are given, Levinson's algorithm of Section 7.2 can be executed by using the following relation between the coefficients of the AR model of order $m-1$ and the coefficients of the AR model of order m

$$a_j^m = a_j^{m-1} - a_m^m a_{m-j}^{m-1}. \tag{7.23}$$

Therefore, if we can estimate the PARCOR a_m^m, the other coefficients can be automatically determined by using the above relation. In Levinson's algorithm, we estimated a_m^m by using the following formula that is obtained by substituting the sample autocovariance functions $\hat{C}_0, \cdots, \hat{C}_m$ into (B.8) in the Appendix B;

$$\begin{aligned}
\hat{a}_m^m &= \left(\hat{C}_0 - \sum_{j=1}^{m-1} \hat{a}_j^{m-1} \hat{C}_j \right)^{-1} \left(\hat{C}_m - \sum_{j=1}^{m-1} \hat{a}_j^{m-1} \hat{C}_{m-j} \right) \\
&= \left(\hat{\sigma}_{m-1}^2 \right)^{-1} \left(\hat{C}_m - \sum_{j=1}^{m-1} \hat{a}_j^{m-1} \hat{C}_{m-j} \right).
\end{aligned} \tag{7.24}$$

In this section, we present another method of estimating PARCOR a_m^m directly from the time series $y_1, \cdots, y_N$ without using the sample autocovariance functions. First of all, let w_n^{m-1} denote the prediction error of the backward AR model with order $m-1$:

$$y_n = \sum_{j=1}^{m-1} a_j^{m-1} y_{n+j} + w_n^{m-1}. \tag{7.25}$$

In the case of a univariate time series, since the autocovariance function is an even function, the AR coefficients of the forward model coincide with those of the backward model. Using this property, from Appendix (B.2), we have the expression

$$\begin{aligned}
C_m - \sum_{j=1}^{m-1} a_j^{m-1} C_{m-j} &= \mathrm{E}\left[\left(y_n - \sum_{j=1}^{m-1} a_j^{m-1} y_{n-j} \right) y_{n-m} \right] \\
&= \mathrm{E}\left[v_n^{m-1} y_{n-m} \right] \\
&= \mathrm{E}\left[v_n^{m-1} w_{n-m}^{m-1} \right].
\end{aligned} \tag{7.26}$$

Therefore the left-hand side of equation (7.26) can be estimated by

$$\frac{1}{N-m} \sum_{n=m+1}^{N} v_n^{m-1} w_{n-m}^{m-1}. \tag{7.27}$$

On the other hand, from (B.4), we have

$$
C_0 - \sum_{j=1}^{m-1} a_j^{m-1} C_j = E\left[\left(y_{n-m} - \sum_{j=1}^{m-1} a_j^{m-1} y_{n-m+j}\right) y_{n-m}\right]
$$
$$
= E\left[w_{n-m}^{m-1} y_{n-m}\right]
$$
$$
= E\left[w_{n-m}^{m-1}\right]^2. \tag{7.28}
$$

Using the equality $E\left(w_{n-m}^{m-1}\right)^2 = E\left(v_n^{m-1}\right)^2$, various estimates of (7.26) can be obtained corresponding to (7.28) as follows;

$$
\frac{1}{N-m} \sum_{n=m+1}^{N} \left(w_{n-m}^{m-1}\right)^2 \tag{7.29}
$$

$$
\frac{1}{N-m}\left\{\sum_{n=m+1}^{N} \left(w_{n-m}^{m-1}\right)^2 \sum_{n=m+1}^{N} \left(v_n^{m-1}\right)^2\right\}^{\frac{1}{2}} \tag{7.30}
$$

$$
\frac{1}{2(N-m)}\left\{\sum_{n=m+1}^{N} \left(w_{n-m}^{m-1}\right)^2 + \sum_{n=m+1}^{N} \left(v_n^{m-1}\right)^2\right\}. \tag{7.31}
$$

Based on these estimates, we obtain the following three estimators of PARCOR

$$
\hat{a}_m^m = \sum_{n=m+1}^{N} v_n^{m-1} w_{n-m}^{m-1}\left\{\sum_{n=m+1}^{N} \left(w_{n-m}^{m-1}\right)^2\right\}^{-1} \tag{7.32}
$$

$$
\hat{a}_m^m = \sum_{n=m+1}^{N} v_n^{m-1} w_{n-m}^{m-1}\left\{\sum_{n=m+1}^{N} \left(w_{n-m}^{m-1}\right)^2 \sum_{n=m+1}^{N} \left(v_n^{m-1}\right)^2\right\}^{-\frac{1}{2}} \tag{7.33}
$$

$$
\hat{a}_m^m = 2\sum_{n=m+1}^{N} v_n^{m-1} w_{n-m}^{m-1}\left\{\sum_{n=m+1}^{N} \left(w_{n-m}^{m-1}\right)^2 + \sum_{n=m+1}^{N} \left(v_n^{m-1}\right)^2\right\}^{-1}. \tag{7.34}
$$

In addition to these estimators, we can define another estimator obtained by replacing $\left(w_{n-m}^{m-1}\right)^2$ with $\left(v_n^{m-1}\right)^2$ in (7.32). The estimate of PARCOR obtained by (7.32) is the regression coefficient when the prediction error v_n^{m-1} of the forward model is regressed on the prediction error w_{n-m}^{m-1} of the backward model. On the other hand, the estimate of (7.33) corresponds to the definition of PARCOR, since it is the correlation coefficient of v_n^{m-1} and w_{n-m}^{m-1}. The estimate of (7.34) minimizes the mean of the variances of the forward prediction errors and the backward prediction errors, and consequently Burg's algorithm based on the maximum entropy method (MEM) is obtained (Burg (1967)).

A generic procedure to estimate the AR model from the time series $y_1, \cdots, y_N$ using the PARCOR method are described below. Here, for simplicity, the mean value of the time series y_n is assumed to be 0.

1. Set $v_n^0 = w_n^0 = y_n$, for $n = 1, \cdots, N$. In addition, for the AR model of order 0, compute $\hat{\sigma}_0^2 = N^{-1} \sum_{n=1}^{N} y_n^2$, and $\text{AIC}_0 = N(\log 2\pi\hat{\sigma}_0^2 + 1) + 2$.

2. For $m = 1, \cdots, M$, repeat the following steps (a)–(f).

    (a) Estimate the PARCOR $\hat{a}_m^m$ by any of the formulae (7.32), (7.33) or (7.34).

    (b) Obtain the AR coefficients $\hat{a}_1^m, \cdots, \hat{a}_{m-1}^m$ by (7.23).

    (c) For $n = m+1, \cdots, N$, obtain the forward prediction error by $v_n^m = v_n^{m-1} - \hat{a}_m^m w_{n-m}^{m-1}$.

    (d) For $n = m+1, \cdots, N$, obtain the backward prediction error by $w_{n-m}^m = w_{n-m}^{m-1} - \hat{a}_m^m v_n^{m-1}$.

    (e) Estimate the innovation variance of the AR model of order m by $\hat{\sigma}_m^2 = \hat{\sigma}_{m-1}^2 \{1 - (\hat{a}_m^m)^2\}$.

    (f) Obtain AIC by $\text{AIC}_m = N(\log 2\pi\hat{\sigma}_m^2 + 1) + 2(m+1)$.

7.5   Large Sample Distribution of the Estimates

On the assumption that the time series is generated by an AR model of order m, for large sample size n, the distribution of the estimates of the AR parameters is approximately given by

$$\hat{a}_j \sim N\left(a_j, n^{-1}\sigma^2\Sigma\right), \tag{7.35}$$

where Σ is the Toepliz matrix (7.2) generated from the autocovariance function and σ^2 is the innovation variance (Brockwell and Davis (1991), Shumway and Stoffer (2000)).

On the other hand, if the time series follows AR model of order m, and if j is larger than m, the estimated PARCOR a_j^j, i.e. the j-th autoregressive coefficient of the AR model of order j ($j > m$), are approximately distributed independently with varaince $1/n$ (Quenouille (1948), Box and Jenkins (1970), Shumway and Stoffer (2000)), i.e.,

$$Var\left(\hat{a}_j^j\right) \simeq \frac{1}{n} \quad \text{for } j > m. \tag{7.36}$$

This property can be used to check the adequacy of the estimated order of the model. The relation between AIC and the estimated PARCOR is considered in Problem 1 of this chapter.

Table 7.1 *Innovation variances and AIC values of the AR models of various orders fitted to the sunspot number data.*

m	σ_m^2	AIC_m	m	σ_m^2	AIC_m	m	σ_m^2	AIC_m
0	0.22900	317.05	7	0.06694	46.95	14	0.05766	26.45
1	0.09204	108.49	8	0.06573	44.73	15	0.05716	26.47
2	0.07058	49.17	9	0.05984	25.02	16	0.05701	27.84
3	0.06959	47.90	10	0.05829	20.96	17	0.05701	29.84
4	0.06868	46.85	11	0.05793	21.52	18	0.05669	30.53
5	0.06815	47.08	12	0.05780	23.02	19	0.05661	32.21
6	0.06805	48.72	13	0.05766	24.47	20	0.05615	32.32

Example (AR modeling for sunspot number data)   The function `arfit` of the package TSSS fits AR model based on the AIC criterion. The following arguments are requires

> lag: the highest order of AR model. Default is $2\sqrt{N}$, where N is the data length.
> method: estimation method:
> 1: Yule-Walker method.
> 2: Least squares (Householder) method.
> 3: PARCOR method (partial autoregression).
> 4: PARCOR method (PARCOR).
> 5: PARCOR method (Burg's algorithm).

and the outputs from this function are:

> sigma2:   innovation variances.
> maice.order: minimum AIC order.
> aic: AIC of the AR models.
> arcoef: AR coefficients of AR models.
> parcor: PARCORs.
> spec: power spectrum (in log scale) of the AIC best AR model.

```
> data(Sunspot)
> arfit( log10(Sunspot), lag=20, method=1 )
```

Table 7.1 summarizes the results of fitting AR models of orders up to 20 by the Yule-Walker method to the logarithm of the sunspot number data shown in Figure 1.1(b). Figure 7.1(a) shows the estimated PAR-

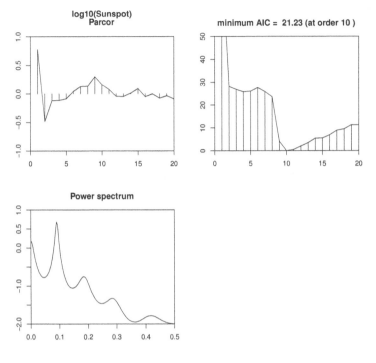

Figure 7.1 *Changes of PARCOR and AIC and estimated spectrum by the AIC best AR model for the sunspot number data.*

COR's for orders $1, \ldots, 20$. From Figure 7.1(b) that shows the change in AIC values as the order m varies, it can be seen that AIC is minimized at $m = 10$, and for larger m it gradually increases.

Since the sample size of the sunspot number data is $n = 231$, from (7.36), the large sample standard error of the estimated PARCOR is $(231)^{-1/2} \simeq 0.066$. It can be seen that the PARCOR's for $m=11, \ldots, 16$ are very small compared with this standard error, which supports the order selected by AIC.

On the other hand, plot (c) shows the spectrum obtained by the AR model of order $m = 10$ that minimizes the AIC. A strong peak is seen in the vicinity of the frequency $f = 0.1$, corresponding to a cycle of approximately eleven years.

Example (AR modeling and power spectra estimated through AR models)

```
> arfit(HAKUSAN[,1], lag=20)
> arfit(log10(Sunspot), lag =20)
> arfit(Temperature, lag=20)
> arfit(BLSALLFOOD, lag=20)
> arfit(WHARD, lag=20)
> arfit(MYE1F, lag=20)
```

Figure 7.2 shows the PARCOR's, changes of AIC and the power spectra obtained by the AIC best orders, when AR models are fitted to the time series shown in Figure 1.1 by the Yule-Walker method.

7.6 Estimation of Multivariate AR Model by the Yule-Walker Method

In this section, estimation methods for multivariate AR model are shown. Hereafter k denotes the number of variables (or dimensions) of a multivariate time series. The parameters of the multivariate AR model with order m

$$y_n = \sum_{i=1}^{m} A_i^m y_{n-i} + v_n, \qquad v_n \sim N(0, V_m), \qquad (7.37)$$

are the variance covariance matrix V_m of the innovation v_n and the AR coefficient matrices $A_1^m, \cdots, A_m^m$ (Akaike and Nakagawa (1989)).

When a multivariate AR model is given, the cross-covariance function is obtained from (6.28) and (6.29). On the other hand, using these equations, the estimates of the parameters of the multivariate AR model can be obtained through the sample cross-covariance function. For actual computation, similarly to the univariate AR model, there is a computationally efficient algorithm. However, for a multivariate time series, the backward model is different from the forward model. In the case of a univariate time series, the forward AR model coincides with the backward AR model, because the autocovariance function is an even function. But this property is not satisfied by multivariate time series. Therefore, in order to derive an efficient algorithm similar to Levinson's algorithm, in addition to (7.37), we have to consider the backward multivariate AR model

$$y_n = \sum_{i=1}^{m} B_i^m y_{n+i} + u_n, \qquad u_n \sim N(0, U_m), \qquad (7.38)$$

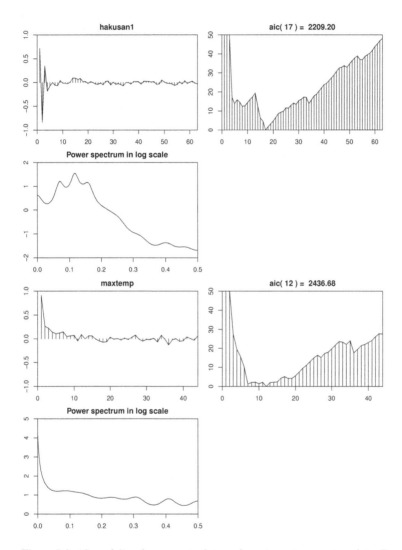

Figure 7.2 *AR modeling for yaw rate data and maximum temperature data. For each data, PARCOR's, change of AIC and estimated power spectrum are shown.*

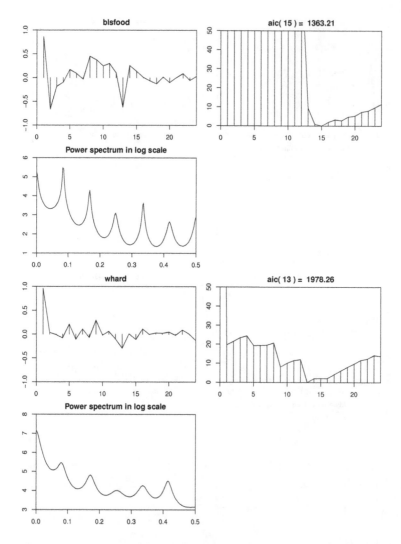

Figure 7.2 *(continued) AR modeling for MYE1F seismic data. PARCOR's, change of AIC and estimated power spectrum are shown.*

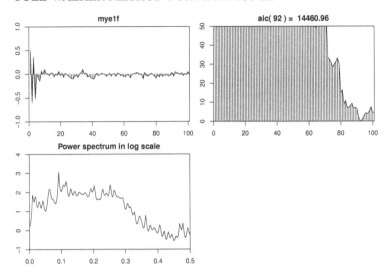

Figure 7.2 *(continued) AR modeling for yaw rate data and maximum tempera-ture data. For each data, PARCOR's, change of AIC and estimated power spec-trum are shown.*

and we need to estimate the variance covariance matrix U_m and the coef-ficients B_i^m, as well as A_i^m and V_m, simultaneously (Whittle (1963)).

1. Set $\hat{V}_0 = \hat{U}_0 = C_0$ and compute the AIC of the AR model of order 0 as

   $$\text{AIC}_0 = N(k\log 2\pi + \log|\hat{V}_0| + k) + k(k+1).$$

2. For $m = 1, \cdots, M$, repeat the following steps (a)–(e).

   (a) $W_m = C_m - \sum_{i=1}^{m-1} A_i^{m-1} C_{m-i}$

   (b) Obtain the PARCOR matrices of the forward and backward AR models by $A_m^m = W_m U_{m-1}^{-1}$ and $B_m^m = W_m^T V_{m-1}^{-1}$.

   (c) Compute the AR coefficients of the forward and backward AR models by $A_i^m = A_i^{m-1} - A_m^m B_{m-i}^{m-1}$ and $B_i^m = B_i^{m-1} - B_m^m A_{m-i}^{m-1}$ for $i = 1, \ldots, m-1$.

   (d) Compute the innovation variance covariance matrices by $V_m = C_0 - \sum_{i=1}^{m} A_i^m C_i^T$ and $U_m = C_0 - \sum_{i=1}^{m} B_i^m C_i$.

   (e) Compute the AIC of the AR model of order m by $\text{AIC}_m = N(k\log 2\pi + \log|\hat{V}_m| + k) + k(k+1) + 2k^2 m.$

By the above-mentioned algorithm, we compute $\text{AIC}_0, \cdots, \text{AIC}_M$, and select the m that attains the minimum of the AICs as the best order of

Table 7.2: *AIC's of multivariate AR models fitted to ship's data.*

m	AIC_m	m	AIC_m	m	AIC_m
0	7091.71	7	5105.83	14	5100.98
1	6238.80	8	5096.35	15	5113.05
2	5275.36	9	5087.91	16	5116.52
3	5173.02	10	5083.79	17	5129.42
4	5135.20	11	5093.79	18	5136.06
5	5136.63	12	5091.42	19	5143.56
6	5121.02	13	5097.98	20	5157.37

the multivariate AR model. In this method, it is assumed that the autoregressive coefficients $a_m(i, j)$ have common orders for all i and j.

Example   The R function `marfit` of package TSSS fits multivariate AR models by the Yule-Walker method and selects the order by minimizing the AIC. The default maximum order is $2\sqrt{N}$ but can be specified the other value by using the parameter `lag`. The outputs from this function are:

`maice.order:`	minimum AIC order.
`aic:`	AIC's of the AR models with orders $0, \dots,$lag.
`arcoef:`	AR coefficients of AR models.
`v:`	innovation variance covariance matrix of the AIC best model.

```
> # Yaw rate, Pitching, Rudder angle
> data(HAKUSAN)
> yy <- as.matrix(HAKUSAN[, c(1,2,4)])
> nc <- dim(yy)[1]
> n <- seq(1, nc, by = 2)
> y <- yy[n, ]
> marfit(y, lag=20)
```

Table 7.2 shows the results of fitting three-variate AR models of orders up to twenty by the Yule-Walker method for the HAKUSAN ship's data shown in Figure 1.1. The AIC is minimized at $m = 10$ and increases gradually afterwards. The power spectra, the cross-spectra, the coherency and the noise contribution, introduced in Chapter 6, are obtained from the estimated multivariate AR model of order $m = 10$.

7.7 Estimation of Multivariate AR Model by the Least Squares Method

To obtain the least squares estimates of the parameters of a multivariate AR model by the Householder method, we first transform the model (7.37) to the following expression with instantaneous response,

$$y_n = B_0 y_n + \sum_{i=1}^{m} B_i y_{n-i} + w_n, \quad w_n \sim N(0, W). \tag{7.39}$$

(Takanami and Kitagawa (1991), Kitagawa and Gersch (1996)). Here, B_0 is a lower triangular matrix whose components on and above the diagonal are zero, namely

$$B_0 = \begin{bmatrix} 0 & 0 & \cdots & 0 \\ b_0(2,1) & 0 & \cdots & 0 \\ \vdots & \ddots & \ddots & \vdots \\ b_0(k,1) & \cdots & b_0(k,k-1) & 0 \end{bmatrix}. \tag{7.40}$$

The variance covariance matrix W is assumed to be a diagonal matrix

$$W = \begin{bmatrix} \sigma_1^2 & 0 & \cdots & 0 \\ 0 & \sigma_2^2 & \ddots & \vdots \\ \vdots & \ddots & \ddots & 0 \\ 0 & \cdots & 0 & \sigma_k^2 \end{bmatrix}. \tag{7.41}$$

Since the model (7.39) can be expressed as

$$y_n = (I - B_0)^{-1} \sum_{i=1}^{m} B_i y_{n-i} + (I - B_0)^{-1} w_n, \tag{7.42}$$

by putting

$$\begin{aligned} A_i &= (I - B_0)^{-1} B_i \\ V &= (I - B_0)^{-1} W (I - B_0)^{-T}, \end{aligned} \tag{7.43}$$

there is a one-to-one correspondence between the multivariate AR model (7.38) and the multivariate model with instantaneous response given in (7.39). Therefore, if the coefficient matrices $B_0, B_1, \cdots, B_m$ and the variances $\sigma_1^2, \cdots, \sigma_k^2$ of the model (7.39) are estimated, the multivariate AR model is also obtained by (7.43).

The advantage of this method is that we do not need to estimate all of the coefficients simultaneously, since the variance covariance matrix W is a diagonal matrix. Namely, if we denote the coefficient matrix B_i as

$$B_i = \begin{bmatrix} b_i(1,1) & \cdots & b_i(1,k) \\ \vdots & \ddots & \vdots \\ b_i(k,1) & \cdots & b_i(k,k) \end{bmatrix}, \tag{7.44}$$

the coefficients of the k models, that is, $\{b_i(p,q), i = 1, \cdots, m, q = 1, \cdots, k, \sigma_p^2\}$ for $p = 1, \cdots, k$ can be estimated independently.

This method is far more computationally efficient than the method that estimates all of the coefficients at once. To realize the above estimation by Householder method, firstly we construct an $(N-m) \times (km+k)$ matrix

$$X = \begin{bmatrix} y_m^T & \cdots & y_1^T & y_{m+1}^T \\ y_{m+1}^T & \cdots & y_2^T & y_{m+2}^T \\ \vdots & \ddots & \vdots & \vdots \\ y_{N-1}^T & \cdots & y_{N-m}^T & y_N^T \end{bmatrix}, \tag{7.45}$$

and transform it to an upper triangular matrix by the Householder transformation,

$$S = HX = \begin{bmatrix} s_{11} & \cdots & s_{1,km+k} \\ & \ddots & \vdots \\ & & s_{km+k,km+k} \\ \mathbf{O} & & \end{bmatrix}. \tag{7.46}$$

It should be noted here that the $(km+1) \times (km+1)$ upper-left submatrix has all the necessary information to estimate the following model for the first component of the time series:

$$y_n(1) = \sum_{i=1}^{j} b_i(1,1)y_{n-i}(1) + \cdots + \sum_{i=1}^{j} b_i(1,k)y_{n-i}(k) + w_n. \tag{7.47}$$

That is, for $j \le m$, the residual variance and the AIC of the j-th order model are obtained by

$$\hat{\sigma}_j^2(1) = \frac{1}{N-m} \sum_{i=kj+1}^{km+1} s_{i,km+1}^2,$$

$$\mathrm{AIC}_j(1) = (N-m)(\log 2\pi \hat{\sigma}_j^2(1) + 1) + 2(kj+1). \tag{7.48}$$

The regression coefficients $c = (c_{11}, \cdots, c_{k1}, \cdots, c_{1j}, \cdots, c_{kj}) = (b_1(1,1), \cdots, b_1(1,k), \cdots, b_j(1,1), \cdots, b_j(1,k))^T$ are obtained as the solutions of the linear equation

$$
\begin{bmatrix} s_{11} & \cdots & s_{1,kj} \\ & \ddots & \vdots \\ O & & s_{kj,kj} \end{bmatrix} \begin{bmatrix} c_{11} \\ \vdots \\ c_{kj} \end{bmatrix} = \begin{bmatrix} s_{1,km+1} \\ \vdots \\ s_{kj,km+1} \end{bmatrix}. \tag{7.49}
$$

The solutions are easily obtained by the following backward substitution:

$$
\begin{aligned}
\hat{c}_{kj} &= s_{kj,km+1}/s_{kj,kj}, \\
\hat{c}_{i\ell} &= (s_{i\ell,km+1} - s_{i\ell,i\ell+1}\hat{c}_{i\ell+1} - \cdots - s_{i\ell,kj}\hat{c}_{kj})/s_{\ell i,\ell i} \\
& \qquad \text{for } i = 1, \cdots, k \text{ and } \ell = 1, \cdots, j. \quad (7.50)
\end{aligned}
$$

Secondly, to estimate the model for the second component of the time series, transform the matrix (7.46) to the following form

$$
S = \begin{bmatrix}
s_{11} & \cdots & s_{1,km} & s_{1,km+1} & s_{1,km+2} & \cdots & s_{1,km+k} \\
s_{21} & \cdots & s_{2,km} & & s_{2,km+2} & \cdots & s_{2,km+k} \\
& \ddots & \vdots & & \vdots & & \vdots \\
& & s_{km+1,km} & & s_{km+1,km+2} & \cdots & s_{km+1,km+k} \\
& & & & s_{km+2,km+2} & \cdots & s_{km+2,km+k} \\
& & & & & \ddots & \vdots \\
& & O & & & & s_{km+k,km+k}
\end{bmatrix},
$$

$$\tag{7.51}$$

by an appropriate Householder transformation. Then the upper-left $(km+2) \times (km+2)$ sub-matrix of S contains all the information necessary for the estimation of the model for the second component.

For $j \leq m$, the innovation variance and the AIC of the j-th order model for the second component are obtained by

$$
\begin{aligned}
\hat{\sigma}_j^2(2) &= \frac{1}{N-m} \sum_{i=kj+2}^{km+2} s_{i,km+2}^2 \\
\text{AIC}_j(2) &= (N-m)(\log 2\pi\hat{\sigma}_j^2(2) + 1) + 2(kj+2). \quad (7.52)
\end{aligned}
$$

To obtain the regression coefficients, we define the $(kj+1)$ - dimensional vector c by $c = (b_1(2,1), \cdots, b_1(2,k), \cdots, b_j(2,1), \cdots, b_j(2,k), b_0(2,1))^T$

and then solve the following system of linear equations:

$$
\begin{bmatrix}
s_{11} & \cdots & s_{1,kj} & s_{1,km+1} \\
s_{21} & \cdots & s_{2,kj} \\
 & \ddots & \vdots \\
 & & s_{kj+1,kj}
\end{bmatrix}
\begin{bmatrix}
c_1 \\
c_2 \\
\vdots \\
c_{kj+1}
\end{bmatrix}
=
\begin{bmatrix}
s_{1,km+2} \\
s_{2,km+2} \\
\vdots \\
s_{kj+1,km+2}
\end{bmatrix}.
$$

$$(7.53)$$

Repeating this procedure up to the k-th component of the time series, we obtain the matrix

$$
S =
\begin{bmatrix}
s_{11} & \cdots & s_{1,km} & s_{1,km+1} & \cdots & s_{1,km+k-1} & s_{1,km+k} \\
\vdots & & \vdots & & \ddots & \vdots & \vdots \\
s_{k-1,1} & \cdots & s_{k-1,km} & & & s_{k-1,km+k-1} & s_{k-1,km+k} \\
s_{k1} & \cdots & s_{k,km} & & & & s_{k,km+k} \\
 & \ddots & \vdots & & & & \vdots \\
 & & s_{km+k-1,km} & & & & s_{km+k-1,km+k} \\
 & & & & & & s_{km+k,km+k} \\
 & & & O
\end{bmatrix}
$$

$$(7.54)$$

by an appropriate Householder transformation.

For $j \leq m$, the innovation variance and the AIC of the j-th order model are obtained by

$$
\hat{\sigma}_j^2(k) = \frac{1}{N-m} \sum_{i=kj+k}^{km+k} s_{i,km+k}^2
$$

$$
\mathrm{AIC}_j(k) = (N-m)(\log 2\pi \hat{\sigma}_j^2(k) + 1) + 2(kj+k). \quad (7.55)
$$

The regression coefficients of this model can be obtained by solving the system of linear equations

$$
\begin{bmatrix}
s_{11} & \cdots & s_{1,kj} & s_{1,km+1} & \cdots & s_{1,q-1} \\
\vdots & & \vdots & & \ddots & \vdots \\
s_{k-1,1} & \cdots & s_{k-1,kj} & & & s_{k-1,q-1} \\
s_{k1} & \cdots & s_{k,kj} & & & \\
 & \ddots & \vdots & & & \\
 & & s_{r,kj}
\end{bmatrix}
\begin{bmatrix}
c_1 \\
\vdots \\
c_{k-1} \\
c_k \\
\vdots \\
c_r
\end{bmatrix}
=
\begin{bmatrix}
s_{1,q} \\
\vdots \\
s_{k-1,q} \\
s_{k,q} \\
\vdots \\
s_{r,q}
\end{bmatrix},
$$

$$(7.56)$$

where the vector c is defined by $c = (b_1(k,1), \cdots, b_1(k,k), \cdots, b_j(k,1), \cdots, b_j(k,k), b_0(k,1), \cdots, b_0(k,k-1))^T$, and $q = km+k$ and $r = kj+k-1$.

This least square method has a significant advantage in that we can select a different order for each variable to enable more flexible modeling than the Yule-Walker method. In addition, it is also possible to specify a time-lag for a variable with respect to other variables or to specify a particular coefficient to be zero. A computer program for estimating such a sophisticated model can be found in Akaike et al. (1979) and the function `mulmar` of the R package `timsac`.

The function `marlsq` of package TSSS fits multivariate AR models by the least squares method based on the Householder transformation. The highest order of the AR model can be specified by the argument `lag`. If it is not given, the default value is $2\sqrt{N}$ where N is the data length. The outputs from this function are:

`maice.order:`	minimum AIC order.
`aic:`	AIC's of the AR models with orders $0, \ldots, $ `lag`.
`arcoef:`	AR coefficients of AR models.
`v:`	innovation variance covariance matrix of the AIC best model.

```
> y <- as.matrix( HAKUSAN )
> z <- marlsq( y, lag=10 )
> z
> marspc(z$arcoef, v = z$v)
```

Example

Multivariate AR model was estimated for four-variate HAKUSAN data consisted of yaw rate, rolling, pitching and rudder angle by putting `lag=10`. The minimum AIC, 6575.741 was attained at $m = 10$. Using the estimated AR coefficient matrices `arcoef` and the innovation variance covariance matrix v, the cross-spectrum, coherency and the power contribution are obtained by the function `marspc`. Figure 7.3 shows the power contribution obtained from the estimated multivariate AR model.

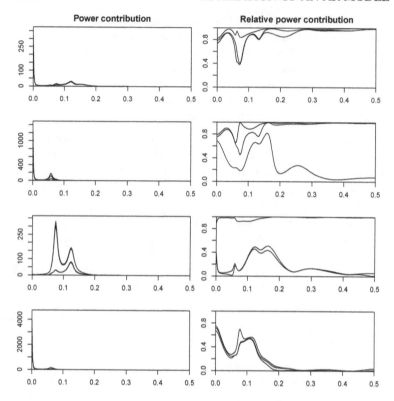

Figure 7.3 *Power contribution of four-variate HAKUSAN data obtained through a multivariate AR model. From top to bottom, yawa rate, rolling, pitching and rudder angle.*

Problems

1. Consider an AR model of order m; $y_n = a_1^m y_{n-1} + \cdots + a_m^m y_{n-m} + v_n$, $v_n \sim N(0, \sigma_m^2)$.

   (1) Using the relation $\sigma_m^2 = (1 - (a_m^m)^2)\sigma_{m-1}^2$, show a criterion to judge whether or not $AR(m)$ is better than $AR(m-1)$.

   (2) The PARCOR coefficients (a_j^j of $AR(j)$) are estimated by using 100 observations, and are given by $a_1^1 = 0.9$, $a_2^2 = -0.6$, $a_3^3 = 0.3$, $a_4^4 = -0.1$, $a_5^5 = 0.15$. Assuming that $C_0 = 1$, compute σ_m^2, for $m = 1, \ldots, 5$.

   (3) Assuming the situation of (2), compute AIC_m for $m = 1, \cdots, 5$ and determine the best order.

2. State the differences in the properties of the Yule-Walker method, the least squares method and the PARCOR method.

3. Show a method, based on AR models, of judging whether or not two time series x_n and y_n are independent.

Chapter 8

The Locally Stationary AR Model

Records of real-world phenomena can mostly be categorized as nonstationary time series. The simplest approach to modeling nonstationary time series is to partition the time interval into several subintervals of appropriate size, on the assumption that the time series are stationary on each subinterval. Then, by fitting an AR model to each subinterval, we can obtain a series of models that approximate nonstationary time series. In this chapter, two modeling methods are shown for the analysis of nonstationary time series, namely, a model for roughly deciding on the number of subintervals and the locations of their endpoints and a model for precisely estimating the change point. A more sophisticated time-varying coefficient AR model will be considered in Chapter 13.

8.1 Locally Stationary AR Model

It is assumed that the time series $y_1, \cdots, y_N$ is nonstationary as a whole, but that we can consider it to be stationary on each subinterval of an appropriately constructed partition. Such a time series that satisfies piecewise stationarity is called a *locally stationary time series* (Ozaki and Tong (1975), Kitagawa and Akaike (1978), Kitagawa and Gersch (1996)). To be specific, k and N_i are the number of subintervals, and the number of observations in the i-th subinterval $(N_1 + \cdots + N_k = N)$, respectively. Actually, k and N_i are unknown in practical modeling. Therefore, in the analysis of locally stationary time series, it is necessary to estimate the number of subintervals, k, the locations of the dividing points and appropriate models for subintervals.

A *locally stationary AR model* is a nonstationary time series model which has the property that, on each appropriately constructed subinterval, it is stationary and can be modeled by an AR model on each of these subintervals. More precisely, consider the i-th subinterval, $[n_{i0}, n_{i1}]$, where

$$n_{i0} = \sum_{j=1}^{i-1} N_j + 1, \qquad n_{i1} = \sum_{j=1}^{i} N_j.$$

For a locally stationary AR model, the time series y_n follows an AR model

$$y_n = \sum_{i=1}^{m_j} a_{ji} y_{n-i} + v_{nj}, \tag{8.1}$$

on the j-th subinterval, where v_{nj} is assumed to be a white noise that satisfies $\mathrm{E}[v_{nj}] = 0$, $\mathrm{E}[v_{nj}^2] = \sigma_j^2$ and $\mathrm{E}[v_{nj} y_{n-m}] = 0$ for $m = 1, 2, \cdots$. The likelihood of the locally stationary AR model is given by

$$L = p(y_1, \cdots, y_N) = \prod_{j=1}^{k} \prod_{n=n_{j0}}^{n_{j1}} p(y_n | y_1, \cdots, y_{n-1}). \tag{8.2}$$

Therefore, similar to the case of the least squares method of the AR model for stationary time series, ignoring the distributions of the first m_1 data points and replacing N_1 by $N_1 - m_1$ and n_{10} with $m_1 + 1$, the likelihood of this model can be approximated by

$$\prod_{j=1}^{k} \left(\frac{1}{2\pi\sigma_j^2} \right)^{\frac{N_j}{2}} \exp\left\{ -\frac{1}{2\sigma_j^2} \sum_{n=n_{j0}}^{n_{j1}} \left(y_n - \sum_{i=1}^{m_j} a_{ji} y_{n-i} \right)^2 \right\}. \tag{8.3}$$

Here, if we consider this likelihood as a function of the number of subintervals k, the length of the j-th interval N_j, the autoregressive order m_j, the autoregressive coefficients $a_j = (a_{j1}, \cdots, a_{jm_j})^T$ and the variance of the white noise σ_j^2, then the log-likelihood function can be expressed as

$$\ell\,(k, N_j, m_j, a_j, \sigma_j^2; j = 1, \cdots, k)$$
$$= -\frac{1}{2} \sum_{j=1}^{k} \left\{ N_j \log 2\pi\sigma_j^2 + \frac{1}{\sigma_j^2} \sum_{n=n_{j0}}^{n_{j1}} \left(y_n - \sum_{i=1}^{m_j} a_{ji} y_{n-i} \right)^2 \right\}. \tag{8.4}$$

For arbitrarily given autoregressive coefficients a_j, by equating the first derivative of the log-likelihood with respect to σ_j^2 to 0, we obtain the maximum likelihood estimate of the variance σ_j^2 as

$$\hat{\sigma}_j^2 = \frac{1}{N_j} \sum_{n=n_{j0}}^{n_{j1}} \left(y_n - \sum_{i=1}^{m_j} a_{ji} y_{n-i} \right)^2. \tag{8.5}$$

Substituting this into (8.4), the log-likelihood becomes

$$\ell\,(k, N_j, m_j, a_j, \hat{\sigma}_j^2; j = 1, \cdots, k)$$

$$= -\frac{1}{2} \sum_{j=1}^{k} \left(N_j \log\left(2\pi\hat{\sigma}_j^2\right) + N_j \right)$$

$$= -\frac{N - m_1}{2} \left(\log 2\pi + 1\right) - \frac{1}{2} \sum_{j=1}^{k} N_j \log \hat{\sigma}_j^2. \qquad (8.6)$$

Therefore, the maximum likelihood estimates of $a_{j1}, \cdots, a_{jm_j}$ are obtained by minimizing σ_j^2 using the least squares method that was described in Chapter 5. Since the AR model on the j-th interval has m_j AR coefficients and the variance as parameters, the AIC for the locally stationary AR model is given by

$$\text{AIC} = (N - m_1)(\log 2\pi + 1) + \sum_{j=1}^{k} N_j \log \hat{\sigma}_j^2 + 2 \sum_{j=1}^{k} (m_j + 1). \qquad (8.7)$$

The number of subintervals k, the length of the j-th subinterval N_j, and the order of the AR model for the j-th interval m_j are obtained by finding these parameters that achieve the minimum AIC value among possible candidates.

8.2 Automatic Partitioning of the Time Interval into an Arbitrary Number of Subintervals

As shown in the previous section, the best locally stationary AR model can, in principle, be obtained by the least squares method and AIC. However, practically speaking, it would require an enormous amount of computation to find the model that minimizes the AIC by fitting locally stationary AR models for all possible combinations of numbers of subintervals, k, and data lengths, $N_1, \cdots, N_k$. In terms of practice, the following procedure was developed to determine the dividing points of the locally stationary AR model (Ozaki and Tong (1975)). Accordingly, only the points $n_i = iL$ are considered as candidates for dividing points, while the minimum unit L of division has been set beforehand. Then, the dividing points of the locally stationary AR model can automatically be decided by the following procedure.

1. Determine the basic span L and the highest order m of the AR models that are fitted to the subinterval of the length L. Here L is set to an appropriate length so that an AR model of order m can be fitted on an interval of length L.

2. Fit AR models of orders up to m to the time series $y_1, \cdots, y_L$, and compute $\text{AIC}_0(0), \cdots, \text{AIC}_0(m)$ to find $\text{AIC}_0 = \min_j \text{AIC}_0(j)$. Further, set $k = 1$, $n_{10} = m+1$, $n_{11} = L$ and $N_1 = L - m$.

3. Fit AR models with orders up to m to the time series $y_{n_{k1}+1}, \cdots, y_{n_{k1}+L}$ and compute $\text{AIC}_1(0)$, $\cdots$, $\text{AIC}_1(m)$ to set $\text{AIC}_1 = \min_j \text{AIC}_1(j)$. AIC_1 is the AIC of a new model that was obtained under the assumption that the model changed at time $n_{k1}+1$. The AIC of the locally stationary AR model that divides the interval $[n_{k0}, n_{k1}+L]$ into two subintervals $[n_{k0}, n_{k1}]$ and $[n_{k1}+1, n_{k1}+L]$ is given by

$$\text{AIC}_D = \text{AIC}_0 + \text{AIC}_1.$$

This model is called a *divided model*.

4. Considering $y_{n_{k0}}, \cdots, y_{n_{k1}+L}$ to be a stationary interval, fit AR models of orders up to m to compute $\text{AIC}_P(0), \cdots, \text{AIC}_P(m)$, and then put $\text{AIC}_P = \min_j \text{AIC}_P(j)$. On the assumption that the time series on the entire interval $[n_{k0}, n_{k1}+L]$ is stationary, the model is called a *pooled model*.

5. To judge the homogeneity of the two subintervals, compare the AIC_D of the model of step 3 and the AIC_P of the model of step 4.

   (a) If $\text{AIC}_D < \text{AIC}_P$, it is judged that a divided model is better. In this case, $n_{k1}+1$ becomes the initial point of the current subinterval, we put $k \equiv k+1$, $n_{k0} \equiv n_{k-1,1}+1$, $n_{k1} = n_{k-1,1}+L$, $N_k = L$ and $\text{AIC}_0 = \text{AIC}_D$.

   (b) If $\text{AIC}_D \geq \text{AIC}_P$, a pooled model is adopted. In this case, the new subinterval $[n_{k1}+1, n_{k1}+L]$ is merged with the former subinterval, and $[n_{k0}, n_{k1}+L]$ becomes the new current subinterval. Therefore, we put $n_{k1} \equiv n_{k1}+L$, $N_k = N_k+L$, and $\text{AIC}_0 = \text{AIC}_P$.

6. If we have at least L remaining additional data points, we go back to step (3). Otherwise, the number of subintervals is k and $[1, n_{11}]$, $[n_{20}, n_{21}], \cdots, [n_{k0}, N]$ are the stationary subintervals.

In the above procedure, we fit two types of AR models whenever an additional time series of length L is obtained. The process can be efficiently carried out by the method of data augmentation shown in Section 5.4 (Kitagawa and Akaike (1978)). In step 2, we firstly construct an $(L-m) \times (m+1)$ matrix from the initial time series; $y_1, \cdots, y_L$,

$$X_0 = \begin{bmatrix} y_m & \cdots & y_1 & y_{m+1} \\ \vdots & \ddots & \vdots & \vdots \\ y_{L-1} & \cdots & y_{L-m} & y_L \end{bmatrix}, \tag{8.8}$$

and then reduce it to upper triangular form by an appropriate Householder transformation H_0,

$$H_0 X_0 = \begin{bmatrix} S \\ O \end{bmatrix} = \begin{bmatrix} s_{11} & \cdots & s_{1m} & s_{1,m+1} \\ & \ddots & \vdots & \vdots \\ & & s_{mm} & s_{m,m+1} \\ & & & s_{m+1,m+1} \\ & O & & \end{bmatrix}. \qquad (8.9)$$

Then the AIC of the AR model of order j fitted to $y_1, \cdots, y_L$ is obtained by

$$\hat{\sigma}_0^2(j) = \frac{1}{L-m} \sum_{i=j+1}^{m+1} s_{i,m+1}^2 \qquad (8.10)$$

$$\mathrm{AIC}_0(j) = (L-m)\log \hat{\sigma}_0^2(j) + 2(j+1). \qquad (8.11)$$

Here, and hereafter in this chapter, we omit the term $(L-m)(\log 2\pi + 1)$ from AIC, since this term is a constant term, irrelevant for model selection. To execute step 3, we construct an $L \times (m+1)$ matrix from $y_{n_{k1}+1}, \cdots, y_{n_{k1}+L}$

$$X_1 = \begin{bmatrix} y_{n_{k1}} & \cdots & y_{n_{k1}-m+1} & y_{n_{k1}+1} \\ \vdots & \ddots & \vdots & \vdots \\ y_{n_{k1}+L-1} & \cdots & y_{n_{k1}+L-m} & y_{n_{k1}+L} \end{bmatrix}, \qquad (8.12)$$

and reduce it to an upper triangular matrix by a Householder transformation H_1,

$$H_1 X_1 = \begin{bmatrix} R \\ O \end{bmatrix} = \begin{bmatrix} r_{11} & \cdots & r_{1m} & r_{1,m+1} \\ & \ddots & \vdots & \vdots \\ & & r_{mm} & r_{m,m+1} \\ & & & r_{m+1,m+1} \\ & O & & \end{bmatrix}. \qquad (8.13)$$

Here, similar to step 2, the AIC of the AR model of order j fitted to the new time series of length L is obtained by

$$\hat{\sigma}_1^2(j) = \frac{1}{L} \sum_{i=j+1}^{m+1} r_{i,m+1}^2 \qquad (8.14)$$

$$\mathrm{AIC}_1(j) = L\log \hat{\sigma}_1^2(j) + 2(j+1). \qquad (8.15)$$

Then,

$$\text{AIC}_D \equiv \min_j \text{AIC}_0(j) + \min_j \text{AIC}_1(j) \qquad (8.16)$$

becomes the AIC for the locally stationary AR model. It is assumed that there was structural change at time $n_{k1} + 1$.

Next, in order to fit an AR model to the pooled data $y_{n_{k0}}, \cdots, y_{n_{k1}+L}$ in step 4, we construct the following $2(m+1) \times (m+1)$ matrix by augmenting the upper triangular matrix S obtained from the data $y_{n_{k0}}, \cdots, y_{n_{k1}}$ with the upper triangular matrix R obtained in step 3,

$$X_2 = \begin{bmatrix} S \\ R \end{bmatrix} = \begin{bmatrix} s_{11} & \cdots & s_{1m} & s_{1,m+1} \\ & \ddots & \vdots & \vdots \\ & & s_{mm} & s_{m,m+1} \\ O & & & s_{m+1,m+1} \\ r_{11} & \cdots & r_{1m} & r_{1,m+1} \\ & \ddots & \vdots & \vdots \\ & & r_{mm} & r_{m,m+1} \\ O & & & r_{m+1,m+1} \end{bmatrix} \qquad (8.17)$$

and reduce it to upper triangular form by a Householder transformation:

$$H_2 X_2 = \begin{bmatrix} T \\ O \end{bmatrix} = \begin{bmatrix} t_{11} & \cdots & t_{1m} & t_{1,m+1} \\ & \ddots & \vdots & \vdots \\ & & t_{mm} & t_{m,m+1} \\ & & & t_{m+1,m+1} \\ & & O & \end{bmatrix} . \qquad (8.18)$$

Then the AIC value for the AR model of order j is obtained by

$$\hat{\sigma}_P^2(j) = \frac{1}{N_k + L} \sum_{i=j+1}^{m+1} t_{i,m+1}^2, \qquad (8.19)$$

$$\text{AIC}_P(j) = (N_k + L) \log \hat{\sigma}_P^2(j) + 2(j+1). \qquad (8.20)$$

Therefore, by finding the minimum (over j) of $\text{AIC}_P(j)$, i.e.,

$$\text{AIC}_P \equiv \min_j \text{AIC}_P(j), \qquad (8.21)$$

we obtain the AIC of the AR model, which was obtained under the assumption that the structural change did not occur at time $n_{k1} + 1$.

In step 5, replace the matrix S with the matrix T if $\text{AIC}_D < \text{AIC}_P$ or with the matrix R if $\text{AIC}_D \geq \text{AIC}_P$. Then go back to step 3.

Example (Locally stationary modeling of seismic data)

The function `lsar` of the package TSSS fits locally stationary AR model and automatically divides the time series into several subintervals on which the time series can be considered as stationary. The arguments to be specified are as follows:

`max.arorder`	highest order of AR models
`ns0`	basic local span.

The outputs from this function are:

`model:`	= 1: pooled model is accepted.
	= 2: switched model is accepted.
`ns:`	number of observations of local spans.
`span:`	start points and end points of local spans.
`nf:`	number of frequencies in computing power spectra.
`ms:`	order of switched model.
`sds:`	innovation variance of switched model.
`aics:`	AIC of switched model.
`mp:`	order of pooled model.
`sdp:`	innovation variance of pooled model.
`aicp:`	AIC of pooled model.
`spec:`	local spectrum.

```
> data(MYE1F)
> lsar(MYE1F, max.arorder = 10, ns0 = 150)
```

Figure 8.1 shows the results of fitting a locally stationary AR model to the east-west component of a seismogram ($N = 2600$) with $L = 150$ and $m = 10$ (Takanami and Kitagawa (1991)). The record involves microtremors as the noise and two types of seismic wave; the P-wave and the S-wave. The power spectra shown in the figures are obtained from AR models estimated on the decided stationary subintervals. Structural changes have been detected at 8 points, $n =$310, 610, 760, 910, 1060, 1360, 1660 and 2010. The change around $n = 600$ corresponds to a change in the spectrum and the variance caused by the arrival of the P-wave. The section $n = 610 - 910$ corresponds to the P-wave. Whereas the spectrum during $n = 610 - 760$ contains a single strong periodic component, various periodic components with different periods are intermingled during the latter half of the P-wave, $n = 760 - 1060$. The S-wave

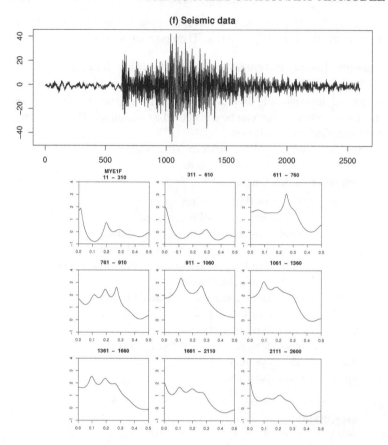

Figure 8.1 *The east-west component record of seismic wave and estimated spectra obtained by a locally stationary AR model.*

appears after $n = 1060$. We can see not only a decrease in power due to the reduction of the amplitude but also that the main peak shifts from the low frequency range to the high frequency range. After $n = 2110$, no significant change in the spectrum could be detected.

8.3   Precise Estimation of the Change Point

In the previous sections, we have presented a method of automatically dividing the time interval of a nonstationary time series into several subintervals in which the time series could be regarded as stationary. Here, we consider a method of detecting the precise time of a structural change by assuming that a structural change of the time series y_n occurred within the time interval $[n_0, n_1]$. A multivariate extension of this method is shown in Takanami and Kitagawa (1991).

Assuming that the structural change occurred at time n; $n_0 \leq n \leq n_1$, a different AR model is fitted to each subinterval $[1, n-1]$ and $[n, N]$, respectively. Then the sum of the two AIC's of the AR models fitted to these time series yields the AIC of a locally stationary AR model with a structural change at time n. To obtain a precise estimate of the time of structural change based on the locally stationary AR models, we compute the AIC's for all n such that $n_0 \leq n \leq n_1$ to find the minimum value. With this method, because we have to estimate AR models for all n, a huge amount of computation is required. However, we can derive a computationally very efficient procedure for obtaining the AIC's for all the locally stationary AR models by using the method of data augmentation shown in Section 5.4.

According to this procedure, from the time series $y_1, \cdots, y_{n_0}$, we first construct an $(n_0 - m) \times (m+1)$ matrix

$$X_0 = \begin{bmatrix} y_m & \cdots & y_1 & y_{m+1} \\ \vdots & \ddots & \vdots & \vdots \\ y_{n_0-1} & \cdots & y_{n_0-m} & y_{n_0} \end{bmatrix}, \tag{8.22}$$

and reduce it to upper triangular form by a Householder transformation:

$$H_0 X_0 = \begin{bmatrix} S \\ O \end{bmatrix} = \begin{bmatrix} s_{11} & \cdots & s_{1m} & s_{1,m+1} \\ & \ddots & \vdots & \vdots \\ & & s_{mm} & s_{m,m+1} \\ & & & s_{m+1,m+1} \\ & & O & \end{bmatrix}. \tag{8.23}$$

Then, the AIC for the AR model of order j fitted to the time series $y_1, \cdots, y_{n_0}$ is obtained by

$$\hat{\sigma}_0^2(j) = \frac{1}{n_0 - m} \sum_{i=j+1}^{m+1} s_{i,m+1}^2, \tag{8.24}$$

$$\text{AIC}_0(j) = (n_0 - m) \log \hat{\sigma}_0^2(j) + 2(j+1). \tag{8.25}$$

Therefore, on the assumption that a structural change occurred at time $n_0 + 1$, the AIC for the best AR model on the first part of the interval is given by

$$\text{AIC}_0^1 \equiv \min_j \text{AIC}_0(j). \tag{8.26}$$

To obtain the AIC for the AR model fitted to the augmented data $y_1, \cdots, y_{n_0+p}$, where p is the number of additional data points ($p \geq 1$), we construct an $(m+p+1) \times (m+1)$ matrix X_1 by augmenting the upper triangular matrix obtained in the previous step with the new data

$$X_1 = \begin{bmatrix} s_{11} & \cdots & s_{1m} & s_{1,m+1} \\ & \ddots & \vdots & \vdots \\ & & s_{mm} & s_{m,m+1} \\ & & & s_{m+1,m+1} \\ y_{n_0} & \cdots & y_{n_0-m+1} & y_{n_0+1} \\ \vdots & & \vdots & \vdots \\ y_{n_0+p-1} & \cdots & y_{n_0-m+p} & y_{n_0+p} \end{bmatrix}, \tag{8.27}$$

and reduce it to upper triangular form by an appropriately defined Householder transformation H_1:

$$H_1 X_1 = \begin{bmatrix} R \\ O \end{bmatrix} = \begin{bmatrix} r_{11} & \cdots & r_{1m} & r_{1,m+1} \\ & \ddots & \vdots & \vdots \\ & & r_{mm} & r_{m,m+1} \\ & & & r_{m+1,m+1} \\ & & O & \end{bmatrix}. \tag{8.28}$$

The AIC for the AR model of order j fitted to the augmented data $y_1, \cdots, y_{n_0+p}$ is obtained by

$$\hat{\sigma}_1^2(j) = \frac{1}{n_0 - m + p} \sum_{i=j+1}^{m+1} r_{i,m+1}^2,$$

$$\text{AIC}_1(j) = (n_0 - m + p) \log \hat{\sigma}_1^2(j) + 2(j+1). \tag{8.29}$$

Therefore, under the assumption that the structural change occurred at time $n_0 + p + 1$, the AIC for the best AR model for the first-half interval is obtained by

$$\text{AIC}_1^1 \equiv \min_j \text{AIC}_1(j). \tag{8.30}$$

Repeating this procedure, the AIC's for the AR models fitted to the time series $\{y_1, \cdots, y_{n_0}\}, \{y_1, \cdots, y_{n_0+p}\}, \cdots, \{y_1, \cdots, y_{n_1}\}$; i.e. $\text{AIC}_0^1, \text{AIC}_1^1, \cdots, \text{AIC}_\ell^1$ can be obtained.

The AIC's for the AR models after the structural change can similarly be obtained. With respect to the AIC's of the latter-half AR models, we first fit AR models to the data $y_{n_1+1}, \cdots, y_N$ and then augment with p observations successively; i.e., we fit AR models to $\{y_{n_1+1}, \cdots, y_N\}, \{y_{n_1-p+1}, \cdots, y_N\}, \{y_{n_1-2p+1}, \cdots, y_N\}, \cdots, \{y_{n_0+1}, \cdots, y_N\}$ and compute the AIC's for the models, AIC_ℓ^2, $\mathrm{AIC}_{\ell-1}^2, \cdots, \mathrm{AIC}_0^2$.

Then,

$$\mathrm{AIC}_j = \mathrm{AIC}_j^1 + \mathrm{AIC}_j^2 \qquad (8.31)$$

yields the AIC for the locally stationary AR model on the assumption that the structural change occurred at time $n_0 + jp + 1$. Therefore, we can estimate the time point of structural change by finding the j for which the minimum of $\mathrm{AIC}_0, \cdots, \mathrm{AIC}_\ell$ is attained.

The function lsar.chgpt of the package TSSS provides a precise estimate of the time of the structural change of the time series. The required parameters for this function are as follows:

max.arorder	highest order of AR model.
subinterval	a vector of the form c(n0,ne), which specifies start and end points of time interval used for modeling.
candidate	a vector of the form c(n1, n2), which gives the minimum and the maximum for change point.
	n0+2k < n1 < n2+k < ne, (k is max.arorder)

The outputs from this function are:

aic:	AIC's of the AR model fitted on [n1,n2].
aicmin:	minimum AIC.
change.point:	estimated change point.
subint:	original sub-interval data.

```
> data(MYE1F)
> lsar.chgpt(MYE1F, max.arorder = 10, subinterval =
c(200,1000), candidate = c(400,800))
>
> lsar.chgpt(MYE1F, max.arorder = 10, subinterval =
c(600,1400), candidate = c(800,1200))
```

Example (Estimation of the arrival times of P-wave and S-wave)

Figure 8.2 shows the results of precisely examining the change points around $n = 600$ and $n = 1000$, where the substantial changes are seen in Figure 8.1. The top plot shows an enlarged view of the $n = 400 - 800$ part. The first half is considered the microtremors and the latter half is considered the P-wave. In the second plot, the value of AIC obtained by

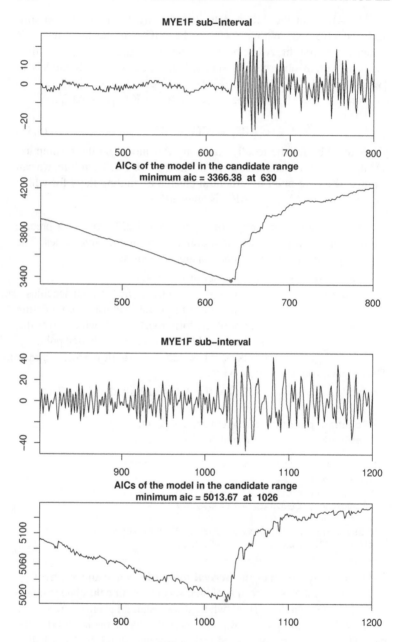

Figure 8.2: *Estimation of the arrival times of the P-wave and the S-wave.*

(8.31) is shown and the minimum value 3366.38 was attained at $n = 630$. Accordingly, it can be inferred that the P-wave arrived at a time corresponding to $n = 630$.

The third plot shows the enlarged view of $n = 800 - 1200$, and the first and the latter half of the plot are the P-wave and the S-wave, respectively. From the value of AIC shown in the fourth plot, we can infer that the S-wave arrived at a time corresponding to $n = 1026$. In the second plot, the AIC has a clear minimum and that indicates the estimate of the arrival time is very accurate. On the other hand, the the fourth plot shows a gradual change in the AIC. It indicates that the detection of the arrival time of S-wave is correspondingly rather more difficult than that of the P-wave.

8.4   Posterior Probability of the Change Point

It is possible to use more fully the information contained in the AIC's. In Akaike (1979), it was shown that $\exp\{-AIC/2\}$ can be considered as an appropriate definition of the likelihood of the model whose parameters are estimated by the maximum likelihood method. In our case

$$p(y|j) = \exp\left\{-\frac{1}{2}AIC_j\right\} \tag{8.32}$$

is the likelihood of the LSAR model, which assumes that $n_0 + pj + 1$ is the arraival time.

Therefore, if the prior distribution of the arrival time is given, then the posterior distribution of the arrival time can be obtained by

$$p(j|y) = \frac{p(y|j)p(j)}{\sum_j p(y|j)p(j)}. \tag{8.33}$$

In the example shown in the next section, the uniform prior over the interval is used. It seems more reasonable to put more weight on the center of the interval. However, since the likelihood, $p(y|j)$, usually takes significant values only at narrow areas, only the local behaviour of the prior is influential to the posterior probability. Therefore as long as very smooth functions are used, the choice of the prior is not so important in the present problem.

One of the most important uses of the estimated arrival time is the determination of the hypocenter of the earthquake. Conventionally, this has been done by the weighted least squares method. However, the use of likelihood of various LSAR models or the posterior distribution of the

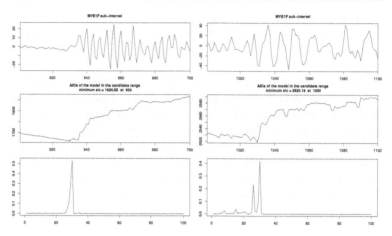

Figure 8.3 *Posterior distribution of the arrival times of the P-wave (left) and the S-wave (right). From top to bottom, seismic data, AIC's and posterior probabilities.*

arrival time would yield a more precise estimate of the hypocenter by using the maximum likelihood method or by a Bayesian modeling.

Using the output of the function lsar.chgpt, the posterior probabilitis of the arrival time can be obtained as follows:

```
> x <- lsar.chgpt(MYE1F, max.arorder=10, subinterval =
c(400,800), candidate = c(600,700))
> AICP <- x$aic
> post <- exp( -(AICP-min(AICP))/2 )
> post <- post/sum(post)
> plot(post,type="l",col="red",lwd=2)
```

Figures 8.3 shows the posterior distrbutions of the arrival times for MYE1F data when the P-wave arrived (left plots) and the S-wave arrived (right plots). In the case of the arrival of the P-wave, the AIC shown in the middle plot has a clear minimum at about 630. On the other hand, in the case of the arrival of the S-wave, change of the AIC is not so significant and actually has several local minima. Note the difference of the scale of vertical axes in the middle plots. As a result, posterior distribution of the P-wave shown in the left plot has a single peak. On the other hand, that of the S-wave has several peaks.

Problems

1. In locally stationary AR modeling, what kind of model should we use if the mean of the time series changes over time. State the expression of the AIC for that model.

2. Corresponding to the time series shown in Figure 1.2, consider a locally stationary AR model for the situation where only the variance of the time series changes over time.

3. Referring to the polynomial regression model introduced in Chapter 11, obtain the AIC of the model in which the polynomial changes over time.

4. In Problem 2, consider models that reflect the continuity or the smoothness of the trend.

5. Assuming that the Householder transformations for (8.22) and (8.27) need $\frac{1}{2}n_0 m^2$ and $\frac{1}{2}(p+1)m^2$ computations, compare the amount of computation required for an ordinary AR model and the locally stationary AR model presented in Section 8.3.

Chapter 9

Analysis of Time Series with a State-Space Model

Various models used in time series analysis can be treated uniformly within the state-space model framework. Many problems of time series analysis can be formulated as the state estimation of a state-space model. This chapter presents the Kalman filter algorithm and the smoothing algorithm for efficient state estimation. In addition, applications to the long-term prediction, interpolation and parameter estimation of a time series are also addressed.

9.1 The State-Space Model

The model for an ℓ-variate time series y_n consisting of the system model and the observation model as follows is called a *state-space model*,

$$
\begin{align}
x_n &= F_n x_{n-1} + G_n v_n, \quad &\text{(system model)} \quad &(9.1) \\
y_n &= H_n x_n + w_n, \quad &\text{(observation model)} \quad &(9.2)
\end{align}
$$

where x_n is a k-dimensional unobservable vector, referred to as the *state* (Anderson and Moore (1979)). v_n is called a system noise or a state noise and is assumed to be an m-dimensional white noise with mean vector zero and variance covariance matrix Q_n. On the other hand, w_n is called observation noise and is assumed to be an ℓ-dimensional Gaussian white noise with mean vector zero and the variance covariance matrix R_n. F_n, G_n, and H_n are $k \times k$, $k \times m$, and $\ell \times k$ matrices, respectively. Many linear models used in time series analysis are expressible in state-space model.

The state-space model includes the following two typical cases. First, if we consider the observation model of (9.2) as a regression model that expresses a mechanism for obtaining the time series y_n, then the state x_n corresponds to the regression coefficients. In this case, the system model (9.1) expresses the time-change of the regression coefficients.

On the other hand, if the state x_n is an unknown signal, the system model expresses the generation mechanism of the signal, and the observation model represents how the actual observations are obtained, namely they are obtained by transforming the signal and adding observation noise.

Example (State-space representation of an AR model)   Consider an AR model for the time series y_n,

$$y_n = \sum_{i=1}^{m} a_i y_{n-i} + v_n, \tag{9.3}$$

where a_i is the AR coefficient and v_n is a Gaussian white noise with mean zero and variance σ^2.

Then, if the state vector is defined as $x_n = (y_n, y_{n-1}, \cdots, y_{n-m+1})^T$, it can easily be verified that there is a relation between the two states, x_n and x_{n-1},

$$x_n = F x_{n-1} + G v_n, \tag{9.4}$$

where F and G are $m \times m$ matrix and m-dimensional vector defined by

$$
F = \begin{bmatrix} a_1 & a_2 & \cdots & a_m \\ 1 & & & \\ & \ddots & & \\ & & 1 & 0 \end{bmatrix}, \qquad
G = \begin{bmatrix} 1 \\ 0 \\ \vdots \\ 0 \end{bmatrix}, \tag{9.5}
$$

respectively. On the other hand, since the first component of the state x_n is the observation y_n, by putting $H = [1\ 0 \cdots 0]$, we obtain the observation model

$$y_n = H x_n. \tag{9.6}$$

Therefore, a state-space model representation of the AR model can be obtained by putting $Q = \sigma^2$ and $R = 0$ to the variances of the system noise and the observation noise, respectively. Thus, the AR model is a special case of the state-space model, in that the state vector x_n is completely determined by the observations until time n, and the observation noise becomes zero.

It should be noted here that the state-space representation of a time series model is not unique, in general. For example, given the state-space models (9.1) and (9.2), for any regular matrix T, by defining a new state z_n, the matrix F_n' and the vectors G_n' and H_n' by

$$z_n = T x_n, \qquad F_n' = T F_n T^{-1}, \qquad G_n' = T G_n, \qquad H_n' = H_n T^{-1}, \tag{9.7}$$

we obtain a state-space model equivalent to the models (9.1) and (9.2)

$$
\begin{aligned}
z_n &= F_n' z_{n-1} + G_n' v_n \\
y_n &= H_n' z_n + w_n.
\end{aligned}
$$

For example, for the AR model (9.3), define a new state vector x_n as

$$
x_n = \left(y_n, \tilde{y}_{n+1|n-1}, \cdots, \tilde{y}_{n+m-1|n-1} \right)^T,
$$

where $\tilde{y}_{n+i|n-1} = \sum_{j=i+1}^{m} a_j y_{n+i-j}$ expresses the part of the one-step-ahead predictor $y_{n+i|n-i+1} = \sum_{j=1}^{m} a_j y_{n+i-j}$ of y_{n+i}, that can be determined from the observations up to time $n-1$. Further, if we define F, G and H by

$$
F = \begin{bmatrix} a_1 & 1 & & \\ a_2 & & \ddots & \\ \vdots & & & 1 \\ a_m & & & 0 \end{bmatrix}, \quad G = \begin{bmatrix} 1 \\ 0 \\ \vdots \\ 0 \end{bmatrix}, \quad H = [1\,0\cdots0], \quad (9.8)
$$

then we obtain another expression of the AR model.

In general, many models treated in this book, such as the ARMA model, the trend component model and the seasonal component model, can be expressed in the form

$$
F_i = \begin{bmatrix} a_{1i} & 1 & & \\ a_{2i} & & \ddots & \\ \vdots & & & 1 \\ a_{mi} & & & 0 \end{bmatrix}, \quad G_i = \begin{bmatrix} 1 \\ b_{1i} \\ \vdots \\ b_{m-1,i} \end{bmatrix}, \quad (9.9)
$$

$$
H_i = [c_{1i},\ c_{2i},\ \cdots,\ c_{m,i}].
$$

In actual time series analysis, a composite model that consists of p components is often used. If the dimensions of the p states are $m_1, \cdots, m_p$, respectively, with $m = m_1 + \cdots + m_p$, then by defining an $m \times m$ matrix, an $m \times p$ matrix, and an m vector

$$
F = \begin{bmatrix} F_1 & & \\ & \ddots & \\ & & F_p \end{bmatrix}, \quad G = \begin{bmatrix} G_1 & & \\ & \ddots & \\ & & G_p \end{bmatrix}, \quad H = [H_1 \cdots H_p],
$$

$$(9.10)$$

a state-space model of the time series is obtained. In this book, this state-space model is used as the standard form.

9.2   State Estimation via the Kalman Filter

A particularly important problem in state-space modeling is to estimate the state x_n based on the observations of the time series y_n. The reason is that tasks such as prediction, interpolation and likelihood computation for the time series can be systematically performed through the state estimation problem.

In this section, we shall consider the problem of estimating the state x_n based on the set of observations $Y_j = \{y_1, \cdots, y_j\}$. In particular, for $j < n$, the state estimation problem results in the estimation of the future state based on present and past observations and is called *prediction*. For $j = n$, the problem is to estimate the current state and is called a *filter*. On the other hand, for $j > n$, the problem is to estimate the past state x_j based on the observations until the present time and is called *smoothing*.

A general approach to these state estimation problems is to obtain the conditional distribution $p(x_n|Y_j)$ of the state x_n given the observations Y_j. Since the state-space model defined by (9.1) and (9.2) is a linear model and the noises v_n and w_n and the initial state x_0 follow normal distributions, all of these conditional distributions become normal distributions. Therefore, to solve the state estimation problem for the state-space model, it is sufficient to obtain the mean vectors and the variance covariance matrices of the conditional distributions.

In general, in order to obtain the conditional joint distribution of states $x_1, \cdots, x_n$ given the observations $y_1, \cdots, y_n$, a huge amount of computation is necessary. However, for the state-space model, a very computationally efficient procedure for obtaining the joint conditional distribution of the state has been developed by means of a recursive computational algorithm. This algorithm is known as the *Kalman filter* (Kalman (1960), Anderson and Moore (1976)). In the following, the conditional mean and the variance covariance matrix of the state x_n are denoted by

$$
\begin{aligned}
x_{n|j} &\equiv \mathrm{E}\left[x_n|Y_j\right] \\
V_{n|j} &\equiv \mathrm{E}\left[(x_n - x_{n|j})(x_n - x_{n|j})^T\right].
\end{aligned}
\tag{9.11}
$$

It is noted that only the conditional distributions with $j = n - 1$ (one-step-ahead prediction) and $j = n$ (filter) are treated in the Kalman filter algorithm. As shown in Figure 9.1, the Kalman filter is realized by repeating the one-step-ahead prediction and the filter with the following algorithm. The derivation of the Kalman filter is shown in Appendix C.

[One-step-ahead prediction]

$$
x_{n|n-1} = F_n x_{n-1|n-1}
$$

$$x_{1|0} \rightarrow x_{2|0} \rightarrow x_{3|0} \rightarrow x_{4|0} \rightarrow x_{5|0} \rightarrow$$
$$\Downarrow$$
$$x_{1|1} \Rightarrow x_{2|1} \rightarrow x_{3|1} \rightarrow x_{4|1} \rightarrow x_{5|1} \rightarrow$$
$$\Downarrow$$
$$x_{1|2} \leftarrow x_{2|2} \Rightarrow x_{3|2} \rightarrow x_{4|2} \rightarrow x_{5|2} \rightarrow$$
$$\Downarrow$$
$$x_{1|3} \leftarrow x_{2|3} \leftarrow x_{3|3} \Rightarrow x_{4|3} \rightarrow x_{5|3} \rightarrow$$
$$\Downarrow$$
$$x_{1|4} \leftarrow x_{2|4} \leftarrow x_{3|4} \leftarrow x_{4|4} \Rightarrow x_{5|4} \rightarrow$$
$$\Downarrow$$

Figure 9.1 *Recursive computation by the Kalman filter and smoothing algorithm.* $\Rightarrow$: *prediction*, $\Downarrow$: *filter*, $\leftarrow$: *smoothing*, $\rightarrow$: *long-term prediction*.

$$V_{n|n-1} = F_n V_{n-1|n-1} F_n^T + G_n Q_n G_n^T. \tag{9.12}$$

[Filter]

$$\begin{aligned}
K_n &= V_{n|n-1} H_n^T \left(H_n V_{n|n-1} H_n^T + R_n \right)^{-1} \\
x_{n|n} &= x_{n|n-1} + K_n(y_n - H_n x_{n|n-1}) \\
V_{n|n} &= (I - K_n H_n) V_{n|n-1}.
\end{aligned} \tag{9.13}$$

In the algorithm for one-step-ahead prediction, the predictor (or mean) vector $x_{n|n-1}$ of x_n is obtained by multiplying the transition matrix F_n by the filter of x_{n-1}, $x_{n-1|n-1}$. Moreover, the variance covariance matrix $V_{n|n-1}$ consists of two terms; the first term expresses the influence of the transformation by F_n and the second expresses the influence of the system noise v_n.

In the filter algorithm, the Kalman gain K_n is initially obtained. The prediction error of y_n and its variance covariance matrices are obtained as $y_n - H_n x_{n|n-1}$ and $H_n V_{n|n-1} H_n^T + R_n$, respectively. Here, the mean vector of the filter of x_n can be obtained as the sum of the predictor $x_{n|n-1}$ and the prediction error multiplied by the Kalman gain. Then, since $x_{n|n}$ can be re-expressed as

$$x_{n|n} = K_n y_n + (I - K_n H_n) x_{n|n-1},$$

it can be seen that $x_{n|n}$ is a weighted sum of the new observation y_n and the predictor, $x_{n|n-1}$.

The variance covariance matrix of the filter $V_{n|n}$ can be written as

$$V_{n|n} = V_{n|n-1} - K_n H_n V_{n|n-1}.$$

Here, the second term of the right-hand side shows the improvement in accuracy of the state estimation of x_n resulting from the information added by the new observation y_n.

9.3   Smoothing Algorithms

The problem of smoothing is to estimate the state vector x_n based on the time series $Y_m = y_1, \cdots, y_m$ for $m > n$. There are three types of smoothing algorithm. If $m = N$, the smoothing algorithm estimates the state based on the entire set of observations and is called *fixed-interval smoothing*. If $n = m - k$, it always estimates the state k steps before, and is called *fixed-lag smoothing*. If n is set to a fixed time point, e.g. $n = 1$, it estimates a specific point such as the initial state and is called *fixed-point smoothing*. Compared with the filtering algorithm that uses the observations up to time n for estimation of the state x_n, fixed-interval smoothing yields a more accurate estimate of the state x_n, by using all available data.

Fixed-interval smoothing

$$
\begin{aligned}
A_n &= V_{n|n} F_{n+1}^T V_{n+1|n}^{-1} \\
x_{n|N} &= x_{n|n} + A_n(x_{n+1|N} - x_{n+1|n}) \\
V_{n|N} &= V_{n|n} + A_n(V_{n+1|N} - V_{n+1|n})A_n^T.
\end{aligned}
\tag{9.14}
$$

As shown in this algorithm, the fixed-interval smoothing estimates, $x_{n|N}$ and $V_{n|N}$, can be derived from results obtained by the Kalman filter, i.e. $x_{n|n-1}$, $x_{n|n}$, $V_{n|n-1}$ and $V_{n|n}$. Therefore, to perform fixed-interval smoothing, we initially obtain $x_{n|n-1}, x_{n|n}, V_{n|n-1}, V_{n|n}$ for $n = 1, \cdots, N$ by the Kalman filter and compute $x_{N-1|N}$, $V_{n-1|N}$ through $x_{1|N}$, $V_{1|N}$ backward in time (see Figure 9.1). It should be noted that the initial values $x_{N|N}$ and $V_{N|N}$ necessary to perform the fixed-interval smoothing algorithm can be obtained by the Kalman filter.

9.4   Long-Term Prediction of the State

It will be shown that by repeating one-step-ahead prediction by the Kalman filter, we can perform long-term prediction, that is, to obtain $x_{n+k|n}$ and $V_{n+k|n}$ for $k = 1, 2, \cdots$. Let us consider the problem of estimating the long-term prediction, i.e. estimating the state x_{n+j} for $j > 1$ based on the time series $Y_n = \{y_1, \cdots, y_n\}$. Firstly, the mean vector $x_{n+1|n}$ and the variance covariance matrix $V_{n+1|n}$ of the one-step-ahead predictor of

x_{n+1} are obtained by the Kalman filter. Here, since the future observation y_{n+1} is unavailable, it is assumed that $Y_{n+1} = Y_n$. In this case, we have that $x_{n+1|n+1} = x_{n+1|n}$ and $V_{n+1|n+1} = V_{n+1|n}$. Therefore, from the one-step-ahead prediction algorithm of the Kalman filter for the period $n+1$, we have

$$
\begin{aligned}
x_{n+2|n} &= F_{n+2}x_{n+1|n} \\
V_{n+2|n} &= F_{n+2}V_{n+1|n}F_{n+2}^T + G_{n+2}Q_{n+2}G_{n+2}^T.
\end{aligned} \tag{9.15}
$$

This means that two-step-ahead prediction can be realized by repeating the prediction step of the Kalman filter twice without the filtering step. In general, j-step-ahead prediction based on Y_n can be performed using the relation that $Y_n = Y_{n+1} = \cdots = Y_{n+j}$, by repeating the prediction step j times. Summarizing the above, the algorithm for the long-term prediction $x_{n+1}, \cdots, x_{n+j}$ based on the observation Y_n can be given as follows:

The long-term prediction
  For $i = 1, \cdots, j$, repeat

$$
\begin{aligned}
x_{n+i|n} &= F_{n+i}x_{n+i-1|n} \\
V_{n+i|n} &= F_{n+i}V_{n+i-1|n}F_{n+i}^T + G_{n+i}Q_{n+i}G_{n+i}^T.
\end{aligned} \tag{9.16}
$$

9.5   Prediction of Time Series

Future values of time series can be immediately predicted by using the predicted state x_n obtained as shown above. When Y_n is given, from the relation between the state x_n and the time series y_n which is expressed by the observation model (9.2), the mean and the variance covariance matrix of y_{n+j} are denoted by $y_{n+j|n} \equiv \mathrm{E}[y_{n+j}|Y_n]$ and $d_{n+j|n} \equiv \mathrm{Cov}(y_{n+j}|Y_n)$, respectively. Then, we can obtain the mean and the variance covariance matrix of the j-step-ahead predictor of the time series y_{n+j} by

$$
\begin{aligned}
y_{n+j|n} &= \mathrm{E}\left[H_{n+j}x_{n+j} + w_{n+j}|Y_n\right] \\
&= H_{n+j}x_{n+j|n} \\
d_{n+j|n} &= \mathrm{Cov}(H_{n+j}x_{n+j} + w_{n+j}|Y_n) \\
&= H_{n+j}\mathrm{Cov}(x_{n+j}|Y_n)H_{n+j}^T + H_{n+j}\mathrm{Cov}(x_{n+j}, w_{n+j}|Y_n) \\
&\quad + \mathrm{Cov}(w_{n+j}, x_{n+j}|Y_n)H_{n+j}^T + \mathrm{Cov}(w_{n+j}|Y_n) \\
&= H_{n+j}V_{n+j|n}H_{n+j}^T + R_{n+j}.
\end{aligned}
$$
$$\tag{9.17}$$
$$\tag{9.18}$$

As indicated previously, the predictive distribution of y_{n+j} based on the observation Y_n of the time series becomes a normal distribution with mean $y_{n+j|n}$ and variance covariance matrix $d_{n+j|n}$. These are easily obtained by (9.17) and (9.18). That is, the mean of the predictor of y_{n+j} is given by $y_{n+j|n}$, and the standard error is given by $(d_{n+j|n})^{1/2}$. It should be noted that the one-step-ahead predictors $y_{n|n-1}$ and $d_{n|n-1}$ of the time series y_n have already been obtained and were applied in the algorithm for the Kalman filter (9.13).

Example (Long-term prediction of BLSALLFOOD data)

The function `tsmooth` of the package TSSS performs long-term prediction by the state-space representation of time series model. The following arguments are required

f:	the state transition matrix F_n.
g:	the matrix G_n.
h:	the matrix H_n.
q:	the system noise covariance matrix Q_n.
r:	the observation noise variance R.
x0:	initial state vector $x_{0|0}$.
v0:	initial state covariance matrix $V_{0|0}$.
filter.end:	end point of filtering.
predict.end:	end pont of prediction.
minmax:	lower and upper limit of observations.

and the outputs from this function are:

mean.smooth:	mean vector of smoother.
cov.smooth:	variance of the smoother.
esterr:	estimation error.
likhood:	log-likelihood.
aic:	AIC of the model.

The function `tsmooth` requires a specific model of time series. In the following code, an AR model was fitted by the function `arfit` and the outputs from the function, i.e. the AIC best AR order, the AR coefficients and the innovation variance are used to define the state-space model. Note that to perform filtering and long-term prediction by using the AR order different from the AIC best order given in the value `z1$maice.order`, set m1 to the desired value, such as `m1 <- 5`.

```
> data(BLSALLFOOD)
> # Fit AR model
> BLS120 <- BLSALLFOOD[1:120]
> z1 <- arfit(BLS120, plot = FALSE)
> tau2 <- z1$sigma2
>
> # Set state-space model
> # m = maice.order, k=1
> m1 <- z1$maice.order
> arcoef <- z1$arcoef[[m1]]
> f <- matrix(0.0e0, m1, m1)
> f[1, ] <- arcoef
> if (m1 != 1)
> for (i in 2:m1) f[i, i-1] <- 1
> g <- c(1, rep(0.0e0, m1-1))
> h <- c(1, rep(0.0e0, m1-1))
> q <- tau2[m1+1]
> r <- 0.0e0
> x0 <- rep(0.0e0, m1)
> v0 <- NULL
>
> # Smoothing by state-space model
> s1 <- tsmooth(BLSALLFOOD, f, g, h, q, r, x0, v0, filter.end =
120, predict.end = 156)
> s1
>
> plot(s1, BLSALLFOOD)
```

Figure 9.2 shows the results of the long-term prediction of the BLSALLFOOD data, $N = 156$. In this prediction, the AR model was fitted to the initial 120 observations, and the estimated AR model was used for long-term prediction of the succeeding 36 observations, $y_{121}, \cdots, y_{156}$. In the estimation of the AR model, we first obtain a time series with mean zero, y_n^*, by deleting the sample mean, $\bar{y}$, of the time series,

$$y_n^* = y_n - \bar{y},$$

and then the parameters of the AR model are obtained by applying the Yule-Walker method to the time series $y_1^*, \cdots, y_N^*$.

The long-term prediction of the time series $y_{n+j|n}^*$ is obtained by applying the Kalman filter to the state-space representation of the AR model; the long-term prediction value of the time series y_{n+j} is then obtained by

$$y_{n+j|n} = y_{n+j|n}^* + \bar{y}.$$

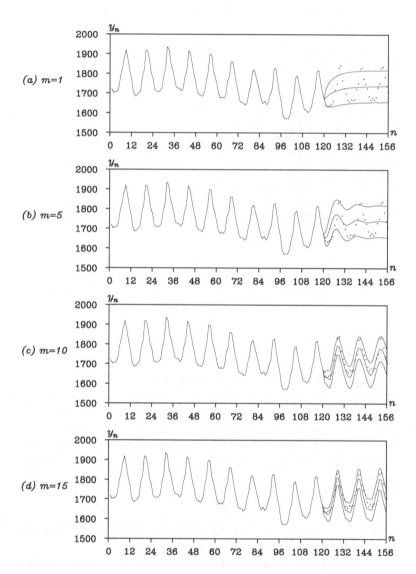

Figure 9.2 *Long-term predictive distributions. (middle line: mean, other two lines: ± (standard deviation) and ○: observed value). Orders of the AR models are 1, 5, 10 and 15, respectively.*

Figure 9.2 shows the mean $y_{120+j|120}$, $j = 1, \cdots, 36$, and its ± 1 standard error interval $y_{120+j|120} \pm \sqrt{d_{120+j|120}}$ of the predictive distribution obtained by this method. The actual time series is indicated by a solid curve for $n \leq 120$ and by the symbol ○ for $n > 120$.

Plots (a), (b), (c) and (d) show the results of the long-term prediction obtained by AR models of orders $m = 1, 5, 10$ and 15, respectively. For the first order AR model shown in plot (a), the long-term prediction value rapidly decays exponentially, which indicates that the information on the periodic behavior of this data is not effectively used for the prediction. In the case of $m = 5$ shown in plot (b), the predictor reasonably reproduced the cyclic behavior for the first year, but after one year passed, the predicted value rapidly decayed. The predictors for the AR model with $m = 10$ reproduce the actual behavior of the time series relatively well. Finally, the predictors for the AR model with $m = 15$ accurately reproduce the details of the wave form of the actual time series. In contrast to the one-step-ahead prediction, the long-term prediction may lead to significant differences among the results from AR models of different assumed orders. These results indicate that prediction by a model of improper order may yield such an inappropriate prediction and that appropriate model selection is crutially important for the long-term prediction.

9.6 Likelihood Computation and Parameter Estimation for Time Series Models

Assume that the state-space representation for a time series model specified by a parameter θ is given. When the time series $y_1, \cdots, y_N$ of length N is given, the N-dimensional joint density function of $y_1, \cdots, y_N$ specified by this time series model is denoted by $f_N(y_1, \cdots, y_N | \theta)$. Then, the likelihood of this model is defined by

$$L(\theta) = f_N(y_1, \cdots, y_N | \theta). \qquad (9.19)$$

By repeatedly applying the relation

$$f_n(y_1, \cdots, y_n | \theta) = f_{n-1}(y_1, \cdots, y_{n-1} | \theta) g_n(y_n | y_1, \cdots, y_{n-1}, \theta),$$

for $n = N, N-1, \cdots, 2$, the likelihood of the time series model can be expressed as a product of conditional density functions:

$$L(\theta) = \prod_{n=1}^{N} g_n(y_n | y_1, \cdots, y_{n-1}, \theta) = \prod_{n=1}^{N} g_n(y_n | Y_{n-1}, \theta). \qquad (9.20)$$

For simplicity of notation, we let $Y_0 = \emptyset$ (empty set) and denote $f_1(y_1|\theta) \equiv g_1(y_1|Y_0, \theta)$. By taking the logarithm of $L(\theta)$, the *log-likelihood* of the model is obtained as

$$\ell(\theta) = \log L(\theta) = \sum_{n=1}^{N} \log g_n(y_n|Y_{n-1}, \theta). \qquad (9.21)$$

As shown in (9.17) and (9.18), since $g_n(y_n|Y_{n-1}, \theta)$ is the conditional distribution of y_n given the observation Y_{n-1} and is a normal distribution with mean $y_{n|n-1}$ and variance covariance matrix $d_{n|n-1}$, it can be expressed as (Kitagawa and Gersch (1996))

$$g_n(y_n|Y_{n-1}, \theta) = \left(\frac{1}{\sqrt{2\pi}}\right)^{\ell} |d_{n|n-1}|^{-\frac{1}{2}} \qquad (9.22)$$
$$\times \exp\left\{-\frac{1}{2}(y_n - y_{n|n-1})^T d_{n|n-1}^{-1}(y_n - y_{n|n-1})\right\}.$$

Therefore, by substituting this density function into (9.21), the log-likelihood of the state-space model is obtained as

$$\ell(\theta) = -\frac{1}{2}\left\{\ell N \log 2\pi + \sum_{n=1}^{N} \log |d_{n|n-1}| \right.$$
$$\left. + \sum_{n=1}^{N}(y_n - y_{n|n-1})^T d_{n|n-1}^{-1}(y_n - y_{n|n-1})\right\}. \qquad (9.23)$$

Stationary time series models such as the AR models, the ARMA models and many other nonstationary time series models such as trend model and seasonal adjustment model can be expressed in the form of linear Gaussian state-space models. Accordingly for such time series models, a unified algorithm for computing the log-likelihood can be obtained by using the Kalman filter and (9.23). The maximum likelihood estimates of the parameters of the time series model can be obtained by maximizing this log-likelihood by a numerical optimization method, which will be described later in Appendix C. Examples of parameter estimation for state-space models are described in Chapter 10 to Chapter 15.

In this way, the parameters contained in the state-space model can be estimated by numerical maximization of the log-likelihood of (9.23), but this generally requires considerable computation. Therefore, if the maximum likelihood estimate or a good approximation can be obtained analytically, that method should be used for efficient estimation. For instance, in estimating an AR model from a time series without missing observations, it is recommended to use either of the Yule-Walker method, the least squares method, or the PARCOR method shown in Chapter 7, instead of the above maximum likelihood method. Furthermore, if an exact maximum likelihood estimate is required, these approximations should be used as initial estimates for numerical optimization.

When maximization of the log-likelihood is necessary but there is not such an approximation method available, we may reduce the dimension of the parameter vector to be estimated by numerical optimization. In the state-space models (9.1) and (9.2), it is assumed that the dimension of the time series is $\ell = 1$ and the variance of w_n is a constant with $R_n = \sigma^2$. Then, if $\tilde{V}_{n|n}$, $\tilde{V}_{n|n-1}$, $\tilde{Q}_n$, and $\tilde{R}$ are defined by

$$V_{n|n-1} = \sigma^2 \tilde{V}_{n|n-1}, \qquad V_{n|n} = \sigma^2 \tilde{V}_{n|n},$$
$$Q_n = \sigma^2 \tilde{Q}_n, \qquad\qquad \tilde{R} = 1, \qquad\qquad (9.24)$$

then it follows that the Kalman filters (9.12) and (9.13) yield identical results, even if those parameters are used.

In one-step-ahead prediction, it is obvious that we can obtain identical results, even if we replace $V_{n-1|n-1}$ and $V_{n|n-1}$ by $\tilde{V}_{n-1|n-1}$ and $\tilde{V}_{n|n-1}$, respectively. In the filtering step, we have

$$
\begin{aligned}
K_n &= V_{n|n-1} H_n^T (H_n V_{n|n-1} H_n^T + R_n)^{-1} \\
&= \sigma^2 \tilde{V}_{n|n-1} H_n^T \sigma^{-2} (H_n \tilde{V}_{n|n-1} H_n^T + 1)^{-1} \\
&= \tilde{V}_{n|n-1} H_n^T (H_n \tilde{V}_{n|n-1} H_n^T + \tilde{R})^{-1} \\
&= \tilde{K}_n. \qquad\qquad\qquad\qquad\qquad\qquad\qquad\qquad (9.25)
\end{aligned}
$$

This indicates that even if we set $\tilde{R} = 1$, the obtained Kalman gain $\tilde{K}_n$ is identical to K_n. Therefore, in the filtering step, we may use $\tilde{V}_{n|n}$ and $\tilde{V}_{n|n-1}$ instead of $V_{n|n}$ and $V_{n|n-1}$. Furthermore, it can be seen that the vectors $x_{n|n-1}$ and $x_{n|n}$ of the state do not change under these modifications. In summary, if R_n is time-invariant and $R = \sigma^2$ is an unknown parameter, we may apply the Kalman filter by setting $R = 1$. Since we then have $d_{n|n-1} = \sigma^2 \tilde{d}_{n|n-1}$ from (9.18), this yields

$$\ell(\theta) = -\frac{1}{2} \left\{ N \log 2\pi\sigma^2 + \sum_{n=1}^{N} \log \tilde{d}_{n|n-1} + \frac{1}{\sigma^2} \sum_{n=1}^{N} \frac{(y_n - y_{n|n-1})^2}{\tilde{d}_{n|n-1}} \right\}.$$
$$(9.26)$$

From the likelihood equation

$$\frac{\partial \ell}{\partial \sigma^2} = -\frac{1}{2} \left\{ \frac{N}{\sigma^2} - \frac{1}{(\sigma^2)^2} \sum_{n=1}^{N} \frac{(y_n - y_{n|n-1})^2}{\tilde{d}_{n|n-1}} \right\} = 0, \qquad (9.27)$$

the maximum likelihood estimate of σ^2 is obtained by

$$\hat{\sigma}^2 = \frac{1}{N} \sum_{n=1}^{N} \frac{(y_n - y_{n|n-1})^2}{\tilde{d}_{n|n-1}}. \qquad (9.28)$$

Furthermore, denoting the parameters in θ except for the variance σ^2 by θ^*, by substituting (9.28) into (9.26), we have

$$\ell(\theta^*) = -\frac{1}{2} \left\{ N \log 2\pi \hat{\sigma}^2 + \sum_{n=1}^{N} \log \tilde{d}_{n|n-1} + N \right\}. \qquad (9.29)$$

By this method, it is possible to reduce the dimension of the parameter vector by one. Summarizing the above, the procedure used here is as follows:

1. Apply the Kalman filter by putting $R = 1$.
2. Obtain an estimate of the variance $\hat{\sigma}^2$ by (9.28).
3. Obtain the log-likelihood $\ell(\theta^*)$ by (9.29).
4. Repeating the above steps (1)–(3), obtain the maximum likelihood estimate $\hat{\theta}^*$ by maximizing the log-likelihood $\ell(\theta^*)$ by means of numerical optimization.

9.7 Interpolation of Missing Observations

In observing time series, a part of the time series might not be able to be obtained due to unexpected factors such as the breakdown of the observational devices, physical constraints of the observed objects or the observation systems. In such cases, the actually unobserved data are called *missing observations* or missing values. Even when only a few percent of the observations are missing, the length of continuously observed data that can be used for the analysis might become very short. In such a situation, we may fill the missing observations with zeros, the mean value of the time series or by linear interpolation. As can be seen in the following numerical examples, such an ad hoc interpolation for missing observations may cause a large bias in the analysis, since it is equivalent to arbitrarily assuming a particular model.

In this section, we shall explain a method of computing the likelihood of the time series model and interpolating the missing observations using the state-space model and the Kalman filter (Jones (1980), Kohn and Ansley (1986), Kitagawa and Gersch (1996)). Using the state-space model of time series, it is possible to compute an exact likelihood even when there are missing observations in the data. Therefore, by maximizing this likelihood, we can obtain maximum likelihood estimates of unknown parameters.

Let $I(n)$ be the set of time instances at which the time series was actually observed. If there are no missing observations, it is obvious that $I(n) = \{1, \cdots, n\}$. Then, for the observations $Y_n \equiv \{y_i \mid i \in I(n)\}$, the log-likelihood of the time series is given by

$$
\begin{aligned}
\ell(\theta) &= \log p(Y_N \mid \theta) \\
&= \sum_{n \in I(N)} \log p(y_n \mid Y_{n-1}, \theta).
\end{aligned}
\tag{9.30}
$$

Since $Y_n \equiv Y_{n-1}$ holds when the observation y_n is missing, as in the case of long-term prediction, the Kalman filter algorithm can be performed by just skipping the filtering step. Namely, the predictive distribution $p(x_n \mid Y_{n-1})$ of the state, or equivalently, the mean $x_{n\mid n-1}$ and the variance covariance matrix $V_{n\mid n-1}$, can be obtained by repeating the prediction step for all n and the filtering step for n such that $n \in I(N)$. Therefore, after computing $y_{n\mid n-1}$ and $d_{n\mid n-1}$ by equations (9.17) and (9.18), similarly to the preceding section, the log-likelihood of the time series model is defined as

$$
\begin{aligned}
\ell(\theta) = -\frac{1}{2} \sum_{n \in I(N)} \Big\{ &\ell \log 2\pi + \log |d_{n\mid n-1}| \\
&+ (y_n - y_{n\mid n-1})^T d_{n\mid n-1}^{-1} (y_n - y_{n\mid n-1}) \Big\}.
\end{aligned}
\tag{9.31}
$$

If a time series model is given, the model can be used for *interpolation of missing observations*. Similar to the likelihood computation, firstly we obtain the predictive distributions $\{x_{n\mid n-1}, V_{n\mid n-1}\}$ and the filter distributions $\{x_{n\mid n}, V_{n\mid n}\}$ using the Kalman filter and skipping the filtering steps when y_n is missing. Then we can obtain the smoothed estimates of the missing observations by applying the fixed-interval smoothing algorithm (9.14). Consequently, the estimate of the missing observation y_n is obtained by $y_{n\mid N} = H_n x_{n\mid N}$ and the variance covariance matrix of the estimate is obtained by $d_{n\mid N} = H_n V_{n\mid N} H_n^T + R_n$.

The results of interpolation by optimal models may be significantly different from conventional interpolations, which are generated
by replacing missing observations with the mean or by straight line interpolation. As can be seen from this example, interpolation of the missing
observations by a particular algorithm corresponds to assuming a particular time series model, different from the true model generating the data.
Therefore, if we perform interpolation without carefully selecting the
model, it may cause significant bad effects in the subsequent analysis.

```
> data(BLSALLFOOD)
> # Fit AR model
> z2 <- arfit(BLSALLFOOD, plot = FALSE)
> tau2 <- z2$sigma2
>
> # m = maice.order, k=1
> m2 <- z2$maice.order
> arcoef <- z2$arcoef[[m2]]
> f <- matrix(0.0e0, m2, m2)
> f[1, ] <- arcoef
> if (m2 != 1)
> for (i in 2:m2) f[i, i-1] <- 1
> g <- c(1, rep(0.0e0, m2-1))
> h <- c(1, rep(0.0e0, m2-1))
> q <- tau2[m2+1]
> r <- 0.0e0
> x0 <- rep(0.0e0, m2)
> v0 <- NULL
>
> tsmooth(BLSALLFOOD, f, g, h, q, r, x0, v0, missed = c(41,
101), np = c(30, 20))
```

Example (Interpolation of missing observations in BLSALLFOOD data)

The function `tsmooth` of the package TSSS can also compute the
log-likelihood of the model for time series with missing observations
and interpolate the missing observations based on the state-space representation of the time series model. Besides the parameters of the state-
space model required for long-term prediction, the information about the
location of the missing observations is necessary. It can be specified by
the arguments `missed` and `np`. Here, the vector `missed` specifies the
start position of the missing observations and the vector `np` specified the
number of continuously missing observations.

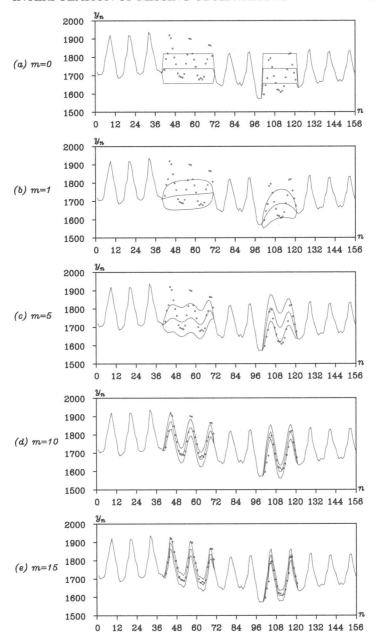

Figure 9.3 *Interpolation of missing values. (middle line: mean, other two lines:* ± *(standard deviation) and* ○: *observed value). Orders of the AR models are 0, 1, 5, 10 and 15.*

In the example below, `missed=c(41,101)` and `np=c(30,20)` indicate that the time series is missing for $n = 41,\ldots,70$ and $n = 101,\ldots,120$. Note that as in the case of long-term prediction, the AR order can be specified by using the argument m2.

Figure 9.3 shows the results of an interpolation experiment for the BLSALLFOOD data. Within 156 observations of data, totally 50 observations, $y_{41},\cdots,y_{70}$ and $y_{101},\cdots,y_{120}$, are assumed to be missing. They are interpolated by AR models of orders 0, 1, 5, 10 and 15, respectively. It should be noted that the order 15 is the AIC best order. AR coefficients are obtained by maximizing the log-likelihood defined by (9.31). Interpolation by the AR model of order 0 is equivalent to replacing the missing observations with the mean of the whole time series. Since the interpolations with the first or the 5th order the AR model cannot capture the wave pattern of the time series well, interpolation with these models cannot reproduce the actual observations shown by the symbol ○. On the other hand, by using the AR model of order 10, the cyclic behavior of the data is reproduced fairly well. Furthermore, it can be seen from plot (e) that the AIC best model of order 15 can reproduce the details of the data remarkably well.

Problems

1. Show that, using the transformation (9.7), we can obtain a state-space model that is equivalent to the one given in (9.1) and (9.2).

2. Assume that the state-space model $x_n = x_{n-1} + v_n$, $y_n = x_n + w_n$ for the one-dimensional state x_n is given, where $v_n \sim N(0, \tau^2)$, $w_n \sim N(0, 1)$ and $x_0 \sim N(0, 10^2)$.
   (1) Show the Kalman filter algorithm for this model.
   (2) Show the relation between $V_{n+1|n}$ and $V_{n|n-1}$.
   (3) If $V_{n|n-1} \to V$ as $n \to \infty$, show that V satisfies the equation $V^2 - \tau^2 V - \tau^2 = 0$.
   (4) Consider the Kalman filter algorithm as $n \to \infty$ (stationary Kalman filter).

3. Let the solution of the equation $V^2 - \tau^2 V - \tau^2 = 0$ be denoted by V. Obtain the variance of the one-step-ahead predictor, filter and smoother in the steady state. For $\tau^2 = 1, 0.1, 0.01, 0.001$, evaluate these variances.

Chapter 10

Estimation of the ARMA Model

In this chapter a method for efficiently computing the log-likelihood of the ARMA model is explained based on the state-space representation and the Kalman filter. Applying the numerical optimization method shown in Chapter 4, it is possible to maximize the log-likelihood. By this procedure, the maximum likelihood estimates of the parameters of the ARMA model can be obtained.

10.1 State-Space Representation of the ARMA Model

Consider a stationary ARMA model (autoregressive moving average model) of order (m, ℓ) (Box and Jenkins (1970), Brockwell and Davis (1981))

$$y_n = \sum_{j=1}^{m} a_j y_{n-j} + v_n - \sum_{j=1}^{\ell} b_j v_{n-j}, \qquad (10.1)$$

where v_n is a Gaussian white noise with mean zero and variance σ^2. Here, a new variable $\tilde{y}_{n+i|n-1}$ is defined as

$$\tilde{y}_{n+i|n-1} = \sum_{j=i+1}^{m} a_j y_{n+i-j} - \sum_{j=i}^{\ell} b_j v_{n+i-j}, \qquad (10.2)$$

which is a part of y_{n+i} that can be directly computable from the observations until time $n-1$, y_{n-1}, y_{n-2}, $\cdots$, and the noise inputs until time n, v_n, v_{n-1}, $\cdots$. Then, the following relations hold.

$$\begin{aligned}
y_n &= a_1 y_{n-1} + \tilde{y}_{n|n-2} + v_n. \\
\tilde{y}_{n+i|n-1} &= a_{i+1} y_{n-1} + \tilde{y}_{n+i|n-2} - b_i v_n. \\
\tilde{y}_{n+k-1|n-1} &= a_k y_{n-1} - b_{k-1} v_n.
\end{aligned} \qquad (10.3)$$

Therefore, setting $k = \max(m, \ell+1)$ and defining the k-dimensional state vector x_n as

$$x_n = (y_n, \tilde{y}_{n+1|n-1}, \cdots, \tilde{y}_{n+k-1|n-1})^T, \qquad (10.4)$$

the ARMA model can be expressed in the form of a state-space model:

$$x_n = Fx_{n-1} + Gv_n$$
$$y_n = Hx_n. \tag{10.5}$$

Here the $k \times k$ matrix F and the k-dimensional vectors G and H are defined as

$$F = \begin{bmatrix} a_1 & 1 & & \\ a_2 & & \ddots & \\ \vdots & & & 1 \\ a_k & & & \end{bmatrix}, \quad G = \begin{bmatrix} 1 \\ -b_1 \\ \vdots \\ -b_{k-1} \end{bmatrix} \tag{10.6}$$

$$H = \begin{bmatrix} 1 & 0 & \cdots & 0 \end{bmatrix},$$

respectively, where $a_i = 0$ for $i > m$ and $b_i = 0$ for $i > \ell$.

In summary, an ARMA model can be expressed by a state-space model where the coefficient matrices F, G and H are time-invariant and the observation noise is zero.

10.2   Initial State Distribution for an AR Model

In order to apply the Kalman filter to the state-space representation of the ARMA model, it is necessary to specify the mean $x_{0|0}$ and the variance covariance matrix $V_{0|0}$ of the initial state. Since $x_{0|0}$ and $V_{0|0}$ express the distribution of the filter without any observations of the time series y_n, they can be evaluated by computing the mean and the variance covariance matrix of the stationary state x_0.

Since the initial variance covariance matrix of an ARMA model is complex, we shall show the one of the AR model in this section. We consider two types of the state-space representation for an AR model. The first is the natural representation shown in (9.3) for which the initial variance covariance matrix has a very simple form. The second respresentation (9.8) is a special form of the state-space representation of an ARMA model, and the initial variance covariance matrix for this representation is generalized to an ARMA model in the next section.

As shown in Chapter 9, an AR model

$$y_n = \sum_{j=1}^{m} a_j y_{n-j} + v_n, \quad v_n \sim N(0, \sigma^2) \tag{10.7}$$

can be expressed in state-space model form,

$$x_n = Fx_{n-1} + Gv_n$$
$$y_n = Hy_n \tag{10.8}$$

by defining the matrices F, G, H and the state vector x_n as follows:

$$F = \begin{bmatrix} a_1 & a_2 & \cdots & a_m \\ 1 & & & \\ & & \ddots & \\ & & & 1 \end{bmatrix}, \quad G = \begin{bmatrix} 1 \\ 0 \\ \vdots \\ 0 \end{bmatrix}, \quad x_n = \begin{bmatrix} y_n \\ y_{n-1} \\ \vdots \\ y_{n-m+1} \end{bmatrix},$$

$$H = [1 \ 0 \ \cdots \ 0]. \tag{10.9}$$

For this state-space representation, the initial state vector is given by $x_0 = [y_0, y_{-1}, \cdots, y_{1-m}]$ and the mean and the variance covariance matrix are obtained by

$$x_{0|0} = E[x_0] = (0, 0, \cdots, 0)^T$$
$$V_{0|0} = \mathrm{Cov}(x_0, x_0) = E[x_0 x_0^T]$$

$$= E \begin{bmatrix} y_0 y_0 & y_0 y_{-1} & \cdots & y_0 y_{1-m} \\ y_{0-1} y_0 & y_{-1} y_{-1} & \cdots & y_{-1} y_{1-m} \\ \vdots & \vdots & \ddots & \vdots \\ y_{1-m} y_0 & y_{1-m} y_{-1} & \cdots & y_{1-m} y_{1-m} \end{bmatrix}$$

$$= E \begin{bmatrix} C_0 & C_1 & \cdots & C_{1-m} \\ C_1 & C_0 & \cdots & C_{2-m} \\ \vdots & \vdots & \ddots & \vdots \\ C_{1-m} & C_{1-m} & \cdots & C_0 \end{bmatrix}. \tag{10.10}$$

As shown in (9.8), another state-space respresentation for the AR model is obtained by

$$F = \begin{bmatrix} a_1 & 1 & & \\ a_2 & & \ddots & \\ \vdots & & & 1 \\ a_m & & & \end{bmatrix}, \quad G = \begin{bmatrix} 1 \\ 0 \\ \vdots \\ 0 \end{bmatrix}, \quad x_n = \begin{bmatrix} y_n \\ \tilde{y}_{n+1|n-1} \\ \vdots \\ \tilde{y}_{n+m-1|n-1} \end{bmatrix},$$

$$H = [1 \ 0 \ \cdots \ 0], \tag{10.11}$$

where $\tilde{y}_{n+1|n-1}$ is defined by $\tilde{y}_{n+1|n-1} = \sum_{j=i+1}^{m} a_j y_{n+i-j}$. Note that this respresention is a special form of the one for an ARMA model, the variance covariance matrix obtained below can be generalized to an ARMA model.

For this representation, the inital state vector is given by $x_0 = [y_n, \tilde{y}_{1|-1}, \cdots, \tilde{y}_{m-1|-1}]$ and therefore we have

$$
\begin{aligned}
V_{0|0} &= \text{Cov}(x_0, x_0) = (V_{ij}) \\
V_{11} &= E[y_0, y_0] = C_0 \\
V_{1i} &= V_{i1} = E[y_0 \tilde{y}_{i-1|-1}] = E\left[y_0 \sum_{j=i}^{m} a_j y_{i-1-j}\right] = \sum_{j=i}^{m} a_j C_{j+1-i} \\
V_{ij} &= E[\tilde{y}_{i-1|-1} \tilde{y}_{j-1|-1}] \\
&= \sum_{p=i}^{m} \sum_{q=j}^{m} a_p a_q E[y_{i-1-p} y_{j-1-q}] = \sum_{p=i}^{m} \sum_{q=j}^{m} a_p a_q C_{q-j-p+i}.
\end{aligned}
$$

Therefore, the variance covariance matrix of the initial state $x_{0|0}$ is obtained by

$$
V_{0|0} = \begin{bmatrix}
C_0 & \sum_{j=2}^{m} a_j C_{j-1} & \cdots & a_m C_1 \\
\sum_{j=2}^{m} a_j C_{j-1} & \sum_{p=2}^{m} \sum_{q=2}^{m} a_p a_q C_{q-p} & \cdots & \sum_{p=2}^{m} a_p a_m C_{m-p} \\
\vdots & \vdots & \ddots & \vdots \\
a_m C_1 & \sum_{q=2}^{m} a_m a_q C_{q-m} & \cdots & a_m a_m C_0
\end{bmatrix}. \quad (10.12)
$$

10.3 Initial State Distribution of an ARMA Model

In this section, we obtain the mean and the variance covariance matrix of the initial state vector $x_{0|0}$ for an ARMA model by extending the one shown in equation (10.12)

Firstly, since $E[v_n] = 0$, it is obvious that $E[y_n] = 0$ and $E(\tilde{y}_{n+i|n-1}) = 0$. Since the expected value of x_n becomes 0, the initial state vector is defined as $x_{0|0} = (0, \cdots, 0)^T$. Next, if the (i, j)-th element of the variance covariance matrix $V_{0|0}$ is denoted by V_{ij}, it can be obtained as

$$
\begin{aligned}
V_{11} &= E[y_0 y_0] = C_0 \\
V_{1i} &= V_{i1} = E\left[y_0 \tilde{y}_{i-1|-1}\right] \\
&= E\left[y_0 \left(\sum_{j=i}^{m} a_j y_{i-1-j} - \sum_{j=i-1}^{\ell} b_j v_{i-1-j}\right)\right] \\
&= \sum_{j=i}^{m} a_j C_{j+1-i} - \sum_{j=i-1}^{\ell} b_j g_{j+1-i} \sigma^2
\end{aligned}
$$

$$
\begin{aligned}
V_{ij} &= \mathrm{E}\left[\tilde{y}_{i-1|-1}\tilde{y}_{j-1|-1}\right] \\
&= \sum_{p=i}^{m}\sum_{q=j}^{m}a_p a_q \mathrm{E}\left[y_{i-1-p}y_{j-1-q}\right] - \sum_{p=i}^{m}\sum_{q=j-1}^{\ell}a_p b_q \mathrm{E}\left[y_{i-1-p}v_{j-1-q}\right] \\
&\quad - \sum_{p=i-1}^{\ell}\sum_{q=j}^{m}b_p a_q \mathrm{E}\left[v_{i-1-p}y_{j-1-q}\right] + \sum_{p=i-1}^{\ell}\sum_{q=j-1}^{\ell}b_p b_q \mathrm{E}\left[v_{i-1-p}v_{j-1-q}\right] \\
&= \sum_{p=i}^{m}\sum_{q=j}^{m}a_p a_q C_{q-j-p+i} - \sum_{p=i}^{m}\sum_{q=j-1}^{\ell}a_p b_q g_{q-j-p+i}\sigma^2 \\
&\quad - \sum_{p=i-1}^{\ell}\sum_{q=j}^{m}b_p a_q g_{p-i-q+j}\sigma^2 + \sum_{p=i-1}^{\ell}b_p b_{p+j-i}\sigma^2.
\end{aligned}
\tag{10.13}
$$

Here we note that C_i and g_i are the autocovariance function and the impulse response function, respectively, of the ARMA model which we can obtain by the method shown in Chapter 6.

10.4 Maximum Likelihood Estimates of an ARMA Model

Using the above results for the state-space representation of the ARMA model, we can compute the log-likelihood for the ARMA model specified by the parameter $\theta = (\sigma^2, a_1, \cdots, a_m, b_1, \cdots, b_\ell)^T$. Firstly, the initial state $x_{0|0}$ and its variance covariance matrix $V_{0|0}$ are provided by the method specified in the preceding section. Next, the one-step-ahead prediction of the state $x_{n|n-1}$ and its variance covariance matrix $V_{n|n-1}$ can be computed by the Kalman filter for $n = 1, \cdots, N$. Then, using (9.23), the log-likelihood of the ARMA model can be obtained as

$$
\ell(\theta) = -\frac{N}{2}\log 2\pi - \frac{1}{2}\sum_{n=1}^{N}\log d_{n|n-1} - \frac{1}{2}\sum_{n=1}^{N}\frac{(y_n - y_{n|n-1})^2}{d_{n|n-1}}, \tag{10.14}
$$

where $d_{n|n-1} = HV_{n|n-1}H^T$ and $y_{n|n-1} = Hx_{n|n-1}$ (Jones (1980), Brockwell and Davis (1981)).

The maximum likelihood estimate of the parameter vector θ can be obtained by maximizing the log-likelihood $\ell(\theta)$ using a numerical optimization method. It is noted that the variance σ^2 can easily be obtained without applying a numerical optimization method by the following procedure which is similar to the method shown in Section 9.6.

1. Apply the Kalman filter after setting $\sigma^2 = 1$ and obtain $d_{n|n-1}$ and $y_{n|n-1}$ for $n = 1, \cdots, N$.

2. Obtain the maximum likelihood estimate of σ^2 by $\hat{\sigma}^2 = N^{-1}\sum_{n=1}^{N}(y_n - y_{n|n-1})^2/d_{n|n-1}$.

3. Obtain the log-likelihood function for the AR coefficients $a_1,\cdots,a_m$, and the MA coefficients $b_1,\cdots,b_\ell$ by $\ell'(a_1,\cdots,a_m,b_1,\cdots,b_\ell) \equiv \ell(\hat{\sigma}^2,a_1,\cdots,a_m,b_1,\cdots,b_\ell)$.

The maximum likelihood estimate of $\theta' = (a_1,\cdots,a_m,b_1,\cdots,b_\ell)^T$ can be obtained by applying a numerical optimization method to the log-likelihood function given in the above-mentioned procedure. In actual computation, however, to satisfy the stationarity and invertibility conditions, we usually apply the following transformations of the parameters.

For the condition of stationarity for the AR coefficients $a_1,\cdots,a_m$, PARCOR's $c_1,\cdots,c_m$ should satisfy $-1 < c_i < 1$ for all $i = 1,\cdots,m$. It can be seen that this condition is guaranteed, if the transformed coefficient α_i defined by

$$\alpha_i = \log\left(\frac{1+c_i}{1-c_i}\right), \tag{10.15}$$

satisfies $-\infty < \alpha_i < \infty$ for all $i = 1,\cdots,m$.

Conversely, for arbitrary $(\alpha_1,\cdots,\alpha_m)^T \in R^m$, if c_i is defined by

$$c_i = \frac{e^{\alpha_i} - 1}{e^{\alpha_i} + 1}, \tag{10.16}$$

then it always satisfies $|c_i| < 1$ and the corresponding AR coefficients satisfy the stationarity condition.

On the other hand, to guarantee the invertibility condition for any $(\beta_1,\cdots,\beta_\ell)^T \in R^\ell$, let d_i be defined as

$$d_i = \frac{e^{\beta_i} - 1}{e^{\beta_i} + 1}, \tag{10.17}$$

and formally obtain the corresponding MA coefficients $b_1,\cdots,b_\ell$ by considering $d_1,\cdots,d_\ell$ to be the PARCOR's.

Then for arbitrary $\theta'' = (\alpha_1,\cdots,\alpha_m,\beta_1,\cdots,\beta_\ell)^T \in R^{m+\ell}$, the corresponding ARMA model will always satisfy the stationarity and invertibility conditions. It is noted that if the coefficient needs to satisfy the condition that $|c_i| < C$ for some $0 < C < 1$, we define

$$c_i = \frac{e^{\alpha_i} - 1}{e^{\alpha_i} + 1}C, \tag{10.18}$$

instead of the equation (10.16).

Table 10.1 *AIC values for ARMA models with various orders fitted to the logarithm of the sunspot number data.*

		ℓ					
		0	1	2	3	4	5
	0	317.12	141.38	73.50	58.75	55.22	54.29
	1	105.43	64.72	139.05	55.96	62.69	58.24
	2	43.26	39.44	41.21	36.57	23.82	25.23
m	3	41.50	41.30	110.15	15.01	25.56	21.26
	4	40.70	42.13	27.16	16.95	14.20	23.11
	5	41.14	42.99	18.78	14.42	14.02	23.84

Example (Sunspot number data)

R function `armafit` estimates parameters of the ARMA model by the maximum likelihood method. `ar.order` and `ma.order` must be given. `ar` and `ma` are the initial AR and MA coefficient vectors, respectively. If they are not given, default values are used in the estimation. The outputs from this function are:

`sigma2`:   innovation variance.
`llkhood`:   log-likelihood of the model.
`aic`:   AIC of the model.
`arcoef`:   AR coefficients.
`macoef`:   MA coefficients.

```
> data(Sunspot) # Sunspot number data
> y <- log10(Sunspot) # y = logarithm of sunspot data
>
> # ARMA(6,3) model
> z1 <- armafit(y, ar.order=3, ma.order=3)
```

For the sunspot number data, the output from the function `armafit` when the ARMA order is set to (3,3) is as follows:

```
sigma2              5.80310e-02
log-likelihood     -0.506
aic                 15.013
AR coefficients     2.541    -2.367     0.804
MA coefficients     1.581    -0.511    -0.177
```

Table 10.1 shows the AIC values when 36 ARMA models are fitted to the logarithm of the sunspot number data within the orders $0 \leq m$,

Table 10.2 *Log-likelihood values of the ARMA models fitted to the logarithm of the sunspot number data.*

		ℓ					
		0	1	2	3	4	5
	0	−157.56	−68.69	−33.75	−25.38	−22.61	−21.14
	1	−50.72	−29.36	**−65.52**	−22.98	**−25.34**	**−22.12**
	2	−18.63	−15.72	−15.61	−12.29	−4.91	−4.62
m	3	−16.64	−15.72	**−48.68**	−1.95	**−4.69**	−1.54
	4	−15.35	−15.06	−6.58	−0.47	1.90	**−1.56**
	5	−14.57	−14.50	−1.39	1.79	2.99	**−0.92**

$\ell \leq 5$ by the function `armafit` independently. From Table 10.1, it can be seen that the smallest AIC, 14.02 is attained at the ARMA model with $m = 5$ and $\ell = 4$. On the other hand, Table 10.2 summarizes the values of the log-likelihoods of the fitted ARMA models $\ell(i, j)$, for $0 \leq i, j \leq 5$.

From Table 10.2, the log-likelihood values for the ARMA models denoted in boldface, such as $\ell(1,2)$, $\ell(3,2)$ and $\ell(1,4)$, etc., are apparently too small compared with the values for other surrounding models. Note that the maximum log-likelihood of the ARMA (i, j) should be larger than or equal to those of ARMA $(i-1, j)$ and ARMA $(i, j-1)$ i.e., $\ell(i, j) \geq \max\{\ell(i-1, j), \ell(i, j-1)\}$, (or equivalently, $\text{AIC}(i, j) \geq \max\{\text{AIC}(i-1, j), \text{AIC}(i, j-1)\} + 2$). Therefore, if the above conditions are violated, it means that the log-likelihood of the ARMA(i, j) model did not converge to the global maximum from the default initial values of the parameters.

10.5   Initial Estimates of Parameters

To mitigate the problem mentioned in the previous section, we consider a method of setting initial parameters for the AR and MA coefficients of the ARMA model. For AR models obtained by setting $\ell = 0$, the initial estimates of the AR coefficients can be rather easily obtained by fitting AR models by the Yule-Walker or least squares method.

For general ARMA(i, j) models, a better model can be always obtained by using the coefficients of a model with a larger log-likelihood between the left and above models in Table 10.2 (namely,

Table 10.3 *Log-likelihood values for some ARMA models obtained by the function* armafit2.

		ℓ					
		0	1	2	3	4	5
	0	−157.56	−68.69	−33.75	−25.38	−22.61	−21.14
	1	−50.72	−29.36	−23.88	−22.98	−22.11	−19.26
	2	−18.63	−15.72	−15.61	−12.29	−4.91	−4.62
m	3	−16.75	−15.65	−15.50	−2.33	−1.89	−0.18
	4	−15.35	−15.06	−6.58	−2.33	−0.45	−0.05
	5	−14.57	−14.50	−1.39	0.76	3.62	3.86

ARMA$(i, j-1)$ and ARMA$(i-1, j)$ models) as the initial values for numerical optimization.

R function armafit2 estimates all ARMA models within the orders by the maximum likelihood method using this method of initializing the AR and MA coefficients. The maximum AR and MA orders (ar.order and ma.order) must be given. The initial AR and MA coefficient vectors are automatically set.

The outputs from this function are:

aicmin:	minimum AIC.
maice.order:	minimum AIC ARMA orders.
sigma2:	innovation variance of all models.
llkhood:	log-likelihood's of all models.
aic:	AIC's of all model.
coef:	AR and MA coefficients of all models.

```
> data(Sunspot) # Sunspot number data
> y <- log10(Sunspot) # y = log( sunspot data )

> armafit2(y, ar.order = 5, ma.order = 5)
```

Example (Sunspot number data: continued)

Table 10.3 and 10.4 show the log-likelihoods and the AIC's of the models obtained by using the function armafit2. It can be seen that the log-likelihood's obtained by this function certainly satisfy the conditions that $\ell(i, j) \geq \ell(i-1, j)$ and $\ell(i, j) \geq \ell(i, j-1)$ for all i and j.

Table 10.4 *AIC values for ARMA models with various orders obtained by the function* `armafit2`.

		ℓ					
		0	1	2	3	4	5
	0	317.12	141.38	73.50	58.75	55.22	54.29
	1	105.43	64.72	55.75	55.96	56.21	52.52
	2	43.25	39.44	41.21	36.57	23.82	25.23
m	3	41.50	41.30	43.00	18.67	19.78	18.75
	4	40.70	42.13	27.16	20.66	18.91	20.37
	5	41.14	42.99	18.78	16.48	12.77	14.27

However, it should be also noted that even if the condition $\ell(i, j) \geq \max\{\ell(i-1, j), \ell(i, j-1)\}$ holds for all i and j, it does not guarantee that the estimated parameters actually converge to the global maxima. Since the ARMA model is very flexible and versatile, we often see that if we fit an ARMA model with large AR and MA orders, it has a tendency to adjust to a small peak or trough of the periodogram. Thus the log-likelihood has many local maxima. To avoid using those inappropriate models in estimating ARMA models, it is recommended to draw the power spectrum corresponding to the estimated ARMA model, to make sure that there are no unnatural troughs in the power spectrum.

Problems

1. Using the expression of equation (9.5), obtain the initial variance covariance matrix for the maximum likelihood estimates of the AR model of order m.

2. Describe what we should check to apply the transformation (10.9) to parameter estimation for the ARMA model.

Chapter 11

Estimation of Trends

In economic time series, we frequently face long-lasting increasing or decreasing trends. In this chapter, we initially consider a polynomial regression model to analyze time series with such tendencies. Secondly, we shall introduce a trend model to estimate a flexible trend that we cannot express by a simple parametric model such as a polynomial or trigonometric regression model. The trend component model treated in this chapter, representing stochastic changes of parameters, forms a framework for modeling various types of nonstationary time series that will be introduced in succeeding chapters.

11.1   The Polynomial Trend Model

The WHARD data in Figure 1.1(e) shows a tendency to increase over the entire time domain. Such a long-term tendency often seen in economic time series such as the one depicted in Figure 1.1(e) is called *trend*. Needless to say, estimating the trend of such a time series is a very natural way of capturing the tendency of the time series and predicting its future behavior. However, even in the case of analyzing short-term behavior of a time series with a trend, we usually analyze the time series after removing the trend, because it is not appropriate to directly apply a stationary time series model such as the AR model to the original series. In this section, we shall explain the *polynomial regression model* as a simple tool to estimate the trend of a time series.

For the polynomial regression model, the time series y_n is expressed as the sum of the trend t_n and a residual w_n

$$y_n = t_n + w_n, \tag{11.1}$$

where w_n follows a Gaussian distribution with mean 0 and variance σ^2. It is assumed that the trend component can be expressed as a polynomial

$$t_n \;\; = \;\; a_0 + a_1 x_n + \cdots + a_m x_n^m. \tag{11.2}$$

181

Since the above polynomial regression model is a special case of the regression model, we can easily estimate the model by the method shown in Chapter 5. Namely, to fit a polynomial trend model, we define the j-th explanatory variable as $x_{nj} = x_n^{j-1}$, and construct the matrix X as

$$X = \begin{bmatrix} 1 & x_1 & \cdots & x_1^m & y_1 \\ \vdots & \vdots & & \vdots & \vdots \\ 1 & x_N & \cdots & x_N^m & y_N \end{bmatrix}. \tag{11.3}$$

Then, by reducing the matrix X to an upper triangular form by an appropriate Householder transformation U,

$$UX = \begin{bmatrix} S \\ O \end{bmatrix} = \begin{bmatrix} s_{11} & \cdots & s_{1,m+2} \\ & \ddots & \vdots \\ & & s_{m+2,m+2} \\ & O & \end{bmatrix}, \tag{11.4}$$

the residual variance of the polynomial regression model of order j is obtained as

$$\hat{\sigma}_j^2 = \frac{1}{N} \sum_{i=j+2}^{m+2} s_{i,m+2}^2. \tag{11.5}$$

Since the j-th order model has $j+2$ parameters, i.e., $j+1$ regression coefficients and the variance, the AIC is obtained as

$$\text{AIC}_j = N \log 2\pi \hat{\sigma}_j^2 + N + 2(j+2). \tag{11.6}$$

The AIC best order is then obtained by finding the minimum of the AIC_j. Given the AIC best order j, the maximum likelihood estimates $\hat{a}_0, \cdots, \hat{a}_j$ of the regression coefficients are obtained by solving the system of linear equations

$$\begin{bmatrix} s_{11} & \cdots & s_{1,j+1} \\ & \ddots & \vdots \\ & & s_{j+1,j+1} \end{bmatrix} \begin{bmatrix} a_0 \\ \vdots \\ a_j \end{bmatrix} = \begin{bmatrix} s_{1,m+2} \\ \vdots \\ s_{j+1,m+2} \end{bmatrix}. \tag{11.7}$$

Here, since the matrix S is in upper triangular form, this system of linear equations can be easily solved by backward substitution.

Example (Maximum temperature data)

R function polreg of the package TSSS fits a polynomial regression model to a time series. The parameter order specifies the highest order

Table 11.1 *Maximum temperature data: The residual variances and AIC values of the polynomial regression models.*

j	$\hat{\sigma}_j^2$	AIC$_j$	j	$\hat{\sigma}_j^2$	AIC$_j$
0	402.62	4296.23	5	10.18	2518.95
1	60.09	3373.76	6	9.64	2594.72
2	58.89	3366.02	7	8.97	2461.42
3	33.61	3095.35	8	8.96	2463.09
4	23.74	2928.47	9	8.96	2465.02

of polynomial regression, and the order of regression is determined by AIC. The function yields the following outputs:

order.maice: Minimum AIC trend order.

sigma2: Residual variances of the models with orders $0 \leq m \leq M$.

aic: AICs of the models with orders $0 \leq m \leq M$.

daic: AIC $-$ min AIC.

coef: Regression coefficents $a(i,m)$, $i = 0, ..., m$, $m = 1, ..., M$.

trend: Estimated trend component by the AIC best model.

For example, to draw the figure of AICs, save the output of polreg to z, then the AICs are given in z$aic.

```
> # Highest Temperature Data of Tokyo
> data(Temperature)
> polreg(Temperature, order = 9)
>
> # plot AIC's of polinomial regression models
> z <- polreg(Temperature, order = 9)
> plot( z$aic,type="b",lwd=2)
>
> # Wholesale hardware data
> data(WHARD)
> y <- log10(WHARD)
> polreg(y, order = 15)
```

Table 11.1 summarizes the residual variance $\hat{\sigma}_j^2$ and the AIC$_j$ of the polynomial regression models fitted to the maximum temperature data shown in Figure 1.1(c). As shown in Table 11.1, although the residual variance $\hat{\sigma}_j^2$ decreases monotonically, AIC was minimized at order 7.

Table 11.2 *Log-WHARD data: The residual variances and AIC values of the polynomial regression models.*

j	$\hat{\sigma}_j^2$	AIC_j	j	$\hat{\sigma}_j^2$	AIC_j
0	0.02752	−550.91	8	0.00115	−1027.44
1	0.00163	−986.60	9	0.00112	−1029.21
2	0.00150	−998.22	10	0.00107	−1033.60
3	0.00149	−996.98	11	0.00107	−1031.65
4	0.00147	−997.43	12	0.00106	−1031.51
5	0.00142	−1000.63	13	0.00106	−1029.62
6	0.00123	−1021.24	14	0.00105	−1029.76
7	0.00122	−1019.98	15	0.00102	−1032.34

Similarly, Table 11.2 shows the residual variances and the AICs of the polynomial regression models with orders $0, \ldots, 15$ fitted to the logarithm of the WHARD data. In this example also, the residual variance decreases as the order increases; the AIC was minimized at order 10.

Plots (a) and (b) of Figure 11.1 show the original time series and the trends estimated using the AIC best polynomial regression models. In plot (a) of the maximum temperature data, a gradual and smooth change of temperature is reasonably captured by the estimated trend. Also in the case of the logarithm of the WHARD data, plot (b) shows a smoothly changing trend. In this case, however, we see that the estimated trend changes too rapidly at the start of the series. Moreover, the abrupt drop in sales in 1974 and 1975 is not clearly detected. We shall consider a method to solve these problems later in Section 11.3.

11.2   Trend Component Model – Model for Gradual Changes

The polynomial trend model treated in the previous section can be expressed as

$$y_n \sim N(t_n, \sigma^2). \tag{11.8}$$

For this model, it is assumed that the time series is distributed as a normal distribution with mean value given by the polynomial t_n and a constant variance σ^2.

This type of parametric model can yield a good estimate of the trend, when the actual trend is a polynomial or can be closely approximated

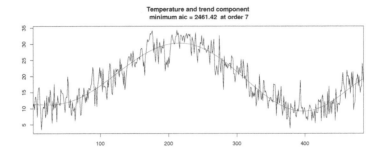

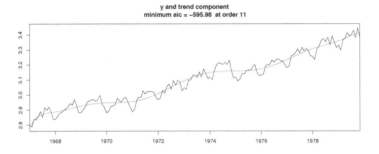

Figure 11.1 *Trends of maximum temperature data and WHARD data estimated by the polynomial regression model.*

by a polynomial. However, in other cases, a parametric model may not reasonably capture the characteristics of the trend, or it may become too sensitive to random noise. Hereafter, we consider a stochastic extension of the polynomial function to mitigate this problem.

Consider a polynomial of the first order, i.e. a straight line, given by

$$t_n = an + b. \tag{11.9}$$

Here, if we define the time difference operator Δ by $\Delta t_n \equiv t_n - t_{n-1}$, then we have

$$\Delta t_n = a, \qquad \Delta^2 t_n = 0. \tag{11.10}$$

This means that the first order polynomial is the solution of the initial value problem for the second order difference equation

$$\Delta^2 t_n = 0, \qquad \Delta t_0 = a, \qquad t_0 = b. \tag{11.11}$$

In general, a polynomial of order $k-1$ can be considered as a solution of the difference equation of order k,

$$\Delta^k t_n = 0. \tag{11.12}$$

In order to make the polynomial more flexible, we assume that $\Delta^k t_n \approx 0$ instead of using the exact difference equation (11.12). This can be achieved by introducing a stochastic difference equation of order k,

$$\Delta^k t_n = v_n, \tag{11.13}$$

where v_n is assumed to be a white noise that follows a normal distribution with mean 0 and variance $\tau^2, N(0, \tau^2)$. In the following, we call the model (11.13) a *trend component model* (Kitagawa and Gersch (1984, 1996)). Because the solution of the difference equation $\Delta^k t_n = 0$ is a polynomial of order $k-1$, the trend component model of order k can be considered as an extension of a polynomial of order $k-1$. When the variance τ^2 of the noise is small, the realization of a trend component model becomes locally a smooth function that resembles the polynomial. However, a remarkable difference from the polynomial is that the trend component model can express a very flexible function globally.

Example (Random walk model)   For $k=1$, this model becomes a *random walk model*, which can be defined by

$$t_n = t_{n-1} + v_n, \quad v_n \sim N(0, \tau^2). \tag{11.14}$$

This model expresses that the trend is locally constant and can be expressed as $t_n \approx t_{n-1}$.

For $k=2$, the trend component model becomes

$$t_n = 2t_{n-1} - t_{n-2} + v_n, \tag{11.15}$$

for which it is assumed that the trend is locally a linear function and satisfies $t_n - 2t_{n-1} + t_{n-2} \approx 0$. In general, the k-th order difference operator of the trend component model (11.13) is given by

$$\Delta^k = (1-B)^k = \sum_{i=0}^{k} {}_k C_i (-B)^i. \tag{11.16}$$

Therefore, using the binomial coefficient $c_i = (-1)^{i+1} {}_k C_i$, the trend component model of order k can be expressed as

$$t_n = \sum_{i=1}^{k} c_i t_{n-i} + v_n. \tag{11.17}$$

Note that $c_1 = 1$ for $k = 1$ and $c_1 = 2$ and $c_2 = -1$ for $k = 2$. Although these models are not stationary, they can formally be considered as AR models of order k. Therefore, by defining the state vector x_n and F, G and H as

$$x_n = \begin{bmatrix} t_n \\ t_{n-1} \\ \vdots \\ t_{n-k+1} \end{bmatrix}, \quad F = \begin{bmatrix} c_1 & c_2 & \cdots & c_k \\ 1 & & & \\ & & \ddots & \\ & & & 1 \end{bmatrix}, \quad G = \begin{bmatrix} 1 \\ 0 \\ \vdots \\ 0 \end{bmatrix},$$
$$H = \begin{bmatrix} 1 & 0 & \cdots & 0 \end{bmatrix}, \tag{11.18}$$

we obtain a state-space representation of the trend component model,

$$\begin{aligned} x_n &= F x_{n-1} + G v_n \\ t_n &= H x_n. \end{aligned} \tag{11.19}$$

Example (State-space representation of trend models)   For $k = 1$, the state-space model is obtained by putting

$$x_n = t_n, \quad F = G = H = 1. \tag{11.20}$$

For $k = 2$, the state-space models of the trend models are obtained by

$$x_n = \begin{bmatrix} t_n \\ t_{n-1} \end{bmatrix}, \quad F = \begin{bmatrix} 2 & -1 \\ 1 & 0 \end{bmatrix}, \quad G = \begin{bmatrix} 1 \\ 0 \end{bmatrix}, \quad H = \begin{bmatrix} 1 & 0 \end{bmatrix} \tag{11.21}$$

or

$$x_n = \begin{bmatrix} t_n \\ -t_{n-1} \end{bmatrix}, \quad F = \begin{bmatrix} 2 & 1 \\ -1 & 0 \end{bmatrix}, \quad G = \begin{bmatrix} 1 \\ 0 \end{bmatrix}, \quad H = \begin{bmatrix} 1 & 0 \end{bmatrix}. \tag{11.22}$$

Note that the model (11.22) is the canonical representation of the trend model treated in Chapter 9. Moreover, for $k = 2$, the state vector can alternatively be defined as $x_n = (t_n, \delta t_n)^T$,

$$F = \begin{bmatrix} 1 & 1 \\ 0 & 1 \end{bmatrix}, \quad G = \begin{bmatrix} 1 \\ 1 \end{bmatrix}, \quad H = \begin{bmatrix} 1 & 0 \end{bmatrix}. \tag{11.23}$$

Since $\delta t_n \equiv \Delta t_n$ holds for this model, we can easily confirm that it is equivalent to the model (11.21). The advantage of using this representation is that we can extend the trend component model by setting (Harvey (1989))

$$v_n = \begin{bmatrix} v_{n1} \\ v_{n2} \end{bmatrix}, \quad F = \begin{bmatrix} 1 & 1 \\ 0 & 1 \end{bmatrix}, \quad G = \begin{bmatrix} 1 & 1 \\ 0 & 1 \end{bmatrix}, \quad H = \begin{bmatrix} 1 & 0 \end{bmatrix}. \tag{11.24}$$

Here, the trend component satisfies

$$
\begin{aligned}
\delta t_n &= \delta t_{n-1} + v_{n2} \\
t_n &= t_{n-1} + \delta t_{n-1} + v_{n1} + v_{n2} \\
&= t_{n-1} + \delta t_n + v_{n1}.
\end{aligned} \tag{11.25}
$$

In contrast to the ordinary trend model with one-dimensional system noise, in this case, we have $\Delta t_n = \delta t_n + v_{n1}$. This extended trend model allows both level and slope to have independent noises, thus expressing a more flexible trend component.

11.3 Trend Model

A trend component in a time series expresses a rough tendency of the phenomenon. In other words, an actually observed time series represents a superposition of a trend and various variations around it. Here, we consider the simplest case where the time series is expressed as

$$
y_n = t_n + w_n, \tag{11.26}
$$

where w_n is a white noise. This is the simplest model that expresses a generating mechanism for observations which can be considered as a special form of the observation model given in Chapter 9.

To estimate the trend component t_n from the time series y_n, we consider the following *trend model* that consists of the trend component model and the above observation model,

$$
\begin{aligned}
\Delta^k t_n &= v_n \tag{11.27} \\
y_n &= t_n + w_n. \tag{11.28}
\end{aligned}
$$

Here, v_n is a Gaussian white noise with mean 0 and variance τ^2 shown in (11.13), and w_n is a Gaussian white noise with mean 0 and variance σ^2.

The observation model in equation (11.28), $y_n = t_n + w_n$, expresses the situation that the time series y_n is obtained by adding an independent noise to the trend. On the other hand, the trend component model (11.27) expresses the change in the trend. Actual time series are usually not as simple as this and often require more sophisticated modeling as will be treated in the next chapter. Based on the state-space representation of the trend component model, the state-space model of the trend is given as follows;

$$
\begin{aligned}
x_n &= F x_{n-1} + G v_n \\
y_n &= H x_n + w_n, \tag{11.29}
\end{aligned}
$$

where the state vector x_n is an appropriately defined k-dimensional vector, and F, G and H are the $k \times k$ matrix, the k-dimensional column vector and the k-dimensional row vector determined by (11.27) and (11.28), respectively. This model differs from the trend component model (11.18) only in that it contains an additional observation noise. As an example, for $k = 2$, the matrices and vectors above are defined by

$$x_n = \begin{bmatrix} t_n \\ t_{n-1} \end{bmatrix}, \quad F = \begin{bmatrix} 2 & -1 \\ 1 & 0 \end{bmatrix}, \quad G = \begin{bmatrix} 1 \\ 0 \end{bmatrix} \quad (11.30)$$
$$H = \begin{bmatrix} 1 & 0 \end{bmatrix}.$$

Once the order k of the trend model and the variances τ^2 and σ^2 have been specified, the smoothed estimates $x_{1|N}, \cdots, x_{N|N}$ are obtained by the Kalman filter and the fixed-interval smoothing algorithm presented in Chapter 9. Since the first component of the state vector is t_n, the first component of $x_{n|N}$, namely $Hx_{n|N}$, is the smoothed estimate of the trend $t_{n|N}$.

Example (Trend of maximum temperature data)

The function trend of the package TSSS fits the trend model and estimates the trend of a time series. The function requires trend.order: the order of the trend component, tau2.ini: the initial estimate of the variance of the system noise, τ^2 and delta: the search width for the τ^2. If tau2=NULL, namely, if an initial estimate of τ^2 is not given, the function yields a rough estimate of τ^2 by searching the smallest AIC within the candidates $\tau^2 = 2^{-k}, k = 1, 2, \dots.$

The function provides the estimated trend (trend), residuals (residual), the variance of the system noise τ^2 (tau2), the variance of the observation noise σ^2 (sigma2), the log-likelihood of the model (llkhood) and the AIC (aic).

```
> # The daily maximum temperatures for Tokyo
> data(Temperature)
> # The first order trend
> trend(Temperature, trend.order = 1, tau2.ini = 0.223, delta =
0.001)
>
> # The second order trend
> trend(Temperature, trend.order = 2)
```

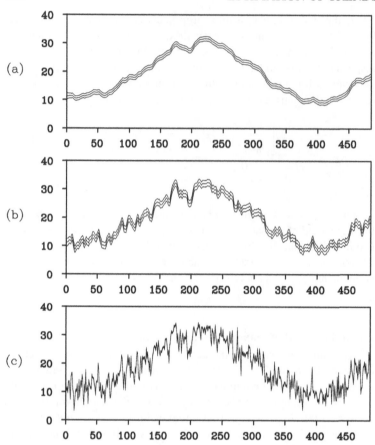

Figure 11.2 *Trend of the temperature data obtained by the first order trend models with different values of τ^2.*

    Figure 11.2 shows various estimates of the trend of the maximum temperature data obtained by changing the variance of the system noise τ^2 for the first order trend model, $k = 1$. The variance of the observation noise σ^2 is estimated by the maximum likelihood method. Plot (a) shows the case of $\tau^2 = 0.223 \times 10^{-2}$. The estimated trend reasonably captures the annual cycles of the temperature data. In plot (b) where the model is $k = 1$ and $\tau^2 = 0.223$, the estimated trend reveals more detailed changes in temperature. Plot (c) shows the case of $k = 1$ and $\tau^2 = 0.223 \times 10^2$. The estimated trend just follows the observed time series.

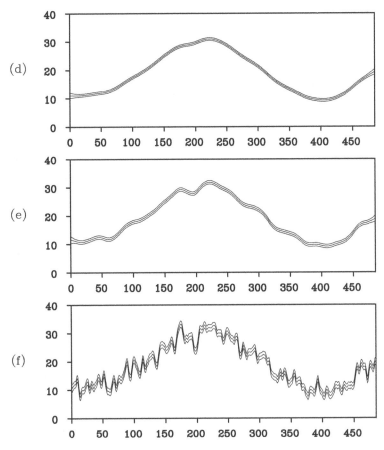

Figure 11.3 *Trend of temperature data obtained by the second order trend models with different values of τ^2.*

On the other hand, Figure 11.3 shows the estimated trends obtained by the models with $k = 2$. The estimated trend in (a) obtained by $\tau^2 = 0.321 \times 10^{-5}$ is too smooth, and the estimated trend in (c) obtained by $\tau^2 = 0.0321$ becomes an undulating curve. But in plot (b), it is evident that the estimate obtained by $\tau^2 = 0.321 \times 10^{-3}$ yields a reasonable trend. Comparing Figures 11.2 (a) and (b) with Figure 11.3 (b), we can see that the second order trend model yields a considerably smoother trend. As shown in the above examples, the trend model contains the order k and the variances τ^2 and σ^2 as parameters, which yield a variety

Table 11.3: *AIC values for trend models.*

τ^2	AIC	τ^2	AIC
	$k=1$		$k=2$
0.223×10^{-2}	2690	0.321×10^{-5}	2556
0.223	2448	0.321×10^{-3}	2506
0.223×10^2	2528	0.0321	2562

of trend estimates. To obtain a good estimate of the trend, it is necessary to select appropriate parameters.

The estimates of the variances τ^2 and σ^2 are obtained by the maximum likelihood method. Using the method shown in Section 9.6, however, if the ratio $\lambda = \tau^2/\sigma^2$ is specified, the estimate of σ^2 is automatically obtained. Therefore, the estimate of the trend is controlled by the variance ratio of the system noise and the observation noise. Here λ is called a *trade-off parameter*. The order of the trend model k can be determined by the information criterion AIC. Often in actual analysis, $k = 2$ is used. However, for situations where the trend is very variable, $k = 1$ might be used instead.

Table 11.3 summarizes the values of τ^2 and AIC for the models used for Figure 11.2. AIC is minimized at $\tau^2 = 0.223$ for $k = 1$ and at $\tau^2 = 0.321 \times 10^{-3}$ for $k = 2$. Incidentally, these estimates are obtained by the maximum likelihood method. Comparison of the AIC values for $k = 1$ and $k = 2$ reveals that the AIC for $k = 1$ is significantly smaller. It should be noted that according to the AIC, the wiggly trend obtained using $k = 1$ is preferable to the nicely smooth trend obtained with $k = 2$. This is probably because the noise w_n is assumed to be a white noise; thus the trend model is inappropriate for the maximum temperature data.

Figure 11.4 shows the observations, the estimated trend and the residuals when the time series is decomposed by the model, $k = 2$ and $\tau^2 = 0.321 \times 10^{-3}$. Obviously, the residuals reveal strong correlation and do not seem to be a white noise sequence.

As shown in the next chapter, we can obtain a smoother trend that fits the data better using a seasonal adjustment model by decomposing the time series into three components; trend, AR component and observation noise. Using a seasonal adjustment model, we can obtain a smoother trend that better fits the data.

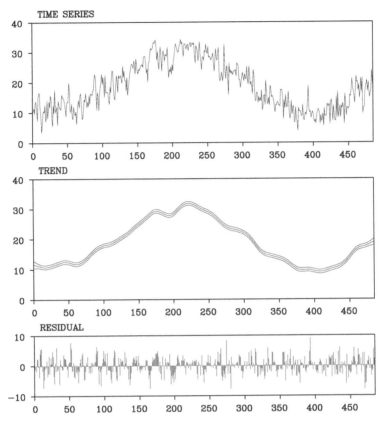

Figure 11.4 *Temperature data and the trend and residuals obtained by the trend model with $k = 2$ and $\tau^2 = 0.321 \times 10^{-3}$.*

Problems

1. According to the random walk hypothesis for stock prices, a stock price y_n follows a random walk model $y_n = y_{n-1} + v_n$, $v_n \sim N(0, \sigma^2)$.

   (1) Assume that $y_n = 17{,}000$ and $\sigma^2 = 40{,}000$ on a certain day. Obtain the k-days ahead prediction of the stock price and its prediction error variance for $k = 1, \ldots, 5$.

   (2) Obtain the probability that the stock price exceeds 17,000 yen after four days have passed.

(3) Do actual stock prices satisfy the random walk hypothesis? If not, consider a modification of the model.

(4) Estimate the trend of actual stock price data. (Nikkei 225 Japanese stock price data is available in the R package TSSS.)

2. Give an example of a parametric trend model other than the polynomial trend model.

Chapter 12

The Seasonal Adjustment Model

In this chapter, the seasonal adjustment model is introduced as an example of extensions of the trend model. Many economic time series repeatedly show similar patterns in the same season of every year. In judging a tendency of a time series or predicting the series, we should carefully take into account these characteristics to avoid misleading results. A seasonal adjustment method is developed for analyzing time series that repeat similar patterns of variation at a constant interval (Shiskin (1976), Akaike (1980), Cleveland et al. (1982), Kitagawa and Gersch (1984)). Seasonal adjustment models decompose a time series y_n into three components of trend t_n, seasonal component s_n and white noise w_n to represent it as $y_n = t_n + s_n + w_n$.

12.1   Seasonal Component Model

In time series, a variable component s_n that repeats every year is called a *seasonal component*. In the following, p denotes the period length of the seasonal component. Here we put $p = 12$ for monthly data and $p = 4$ for quarterly data, respectively. Then, the seasonal component s_n approximately satisfies

$$s_n = s_{n-p}. \tag{12.1}$$

Using the lag operator B, s_{n-p} is denoted as $B^p s_n$, and the seasonal component approximately satisfies

$$(1 - B^p)s_n = 0. \tag{12.2}$$

Similar to the stochastic trend component model introduced in Chapter 11, a model for the seasonal component that gradually changes with time, is given by (Kitagawa and Gersch (1984, 1996))

$$(1 - B^p)^\ell s_n = v_{n2}, \quad v_{n2} \sim N(0, \tau_2^2). \tag{12.3}$$

195

In particular, putting $\ell = 1$, we can obtain a random walk model for the seasonal component by

$$s_n = s_{n-p} + v_{n2}. \tag{12.4}$$

In this model, it is assumed that s_{pn+i}, $n = 1, 2, \ldots$ is a random walk for any $i = 1, \ldots, p$.

Therefore, assuming that the time series consists of the trend component t_n, the seasonal component s_n and the observation noise w_n, we obtain a basic model for seasonal adjustment as

$$y_n = t_n + s_n + w_n, \tag{12.5}$$

with the trend component model (11.15) in the previous chapter and the above seasonal component model (12.3).

However, the apparently most natural model (12.4) for seasonal adjustment would not work well in practice, because the trend component model and the seasonal component model both contain the common factor $(1 - B)^q$, $(q \geq 1)$. This can be seen by comparing the back-shift operator expression of the trend model (11.18) to the seasonal component model (12.3) with the decomposition

$$(1 - B^p)^\ell = (1 - B)^\ell (1 + B + \cdots + B^{p-1})^\ell.$$

Here, assume that e_n is an arbitrary solution of the difference equation

$$(1 - B)^q e_n = 0. \tag{12.6}$$

For $q = 1$, e_n is an arbitrary constant. If we define new components t'_n and s'_n as

$$\begin{aligned} t'_n &= t_n + e_n \\ s'_n &= s_n - e_n, \end{aligned}$$

then they satisfy (11.15), (12.3) and

$$y_n = t'_n + s'_n + w_n. \tag{12.7}$$

Therefore, we have infinitely many ways to decompose the time series yielding the same noise inputs v_{n1}, v_{n2} and w_n. Moreover, since the likelihood of the model corresponding to those decompositions is determined only by v_{n1}, v_{n2} and w_n, it is impossible to discriminate between the goodness of the decompositions by the likelihood. Therefore, if you use component models with common elements, the uniqueness of the decomposition is lost.

A simple way to guarantee uniqueness of decomposition is to ensure that none of the component models share any common factors. Since $1 - B^p = (1 - B)(1 + B + \cdots + B^{p-1})$ and the sufficient condition for $(1 - B^p)^\ell = 0$ is

$$(1 + B + \cdots + B^{p-1})^\ell = 0, \tag{12.8}$$

$S_n \approx S_{n-p}$ is attained if

$$\sum_{i=0}^{p-1} B^i s_n \approx 0 \tag{12.9}$$

is satisfied. Therefore, we will use the following model as a stochastic model for a seasonal component that changes gradually with time:

$$\left(\sum_{i=0}^{p-1} B^i\right)^\ell s_n = v_{n2}, \quad v_{n2} \sim N\left(0, \tau_2^2\right). \tag{12.10}$$

In this book, the above model is called a *seasonal component model* with period p and order ℓ. In actual analysis, except for situations where the seasonal component shows a significant trend in its changes, the first order model

$$\sum_{i=0}^{p-1} s_{n-i} = v_{n2}, \quad v_{n2} \sim N\left(0, \tau_2^2\right) \tag{12.11}$$

is usually used.

To obtain a state-space representation of the seasonal component model, initially we expand the operator in (12.9) as follows:

$$\left(\sum_{i=0}^{p-1} B^i\right)^\ell = 1 - \sum_{i=1}^{\ell(p-1)} d_i B^i. \tag{12.12}$$

The coefficient d_i is given by $d_i = -1, i = 1, \ldots, p-1$ for $\ell = 1$ and $d_i = -i - 1, i \leq p - 1$, and $d_i = i + 1 - 2p, p \leq i \leq 2(p-1)$ for $\ell = 2$. Since the seasonal component model given in (12.12) is formally a special case of an autoregressive model, the state-space representation of the seasonal component model can be given as

$$x_n = \begin{bmatrix} s_n \\ s_{n-1} \\ \vdots \\ s_{n-\ell(p-1)+1} \end{bmatrix}, \quad F = \begin{bmatrix} d_1 & d_2 & \cdots & d_{\ell(p-1)} \\ 1 & & & \\ & \ddots & & \\ & & 1 & \end{bmatrix}, \quad G = \begin{bmatrix} 1 \\ 0 \\ \vdots \\ 0 \end{bmatrix}$$

$$H = [\, 1, 0, \ldots, 0 \,]. \tag{12.13}$$

12.2   Standard Seasonal Adjustment Model

For standard seasonal adjustment, the time series y_n is decomposed into the following three components

$$y_n = t_n + s_n + w_n, \qquad (12.14)$$

where t_n, s_n and w_n are the trend component, the seasonal component and the observation noise, respectively. Combining the basic model (12.12) with the trend component model and the seasonal component model, we obtain the following standard seasonal adjustment models,

$$
\begin{aligned}
y_n &= t_n + s_n + w_n && \text{(observation model)} && (12.15)\\
\Delta^k t_n &= v_{n1} && \text{(trend component model)} && (12.16)\\
\left(\sum_{i=0}^{p-1} B^i \right)^{\ell} s_n &= v_{n2} && \text{(seasonal component model)}, && (12.17)
\end{aligned}
$$

where $w_n \sim N(0, \sigma^2)$, $v_{n1} \sim N(0, \tau_1^2)$ and $v_{n2} \sim N(0, \tau_2^2)$.

The above model is called the standard *seasonal adjustment model*. For a seasonal adjustment model with trend order k, period p and seasonal order $\ell = 1$, let the $(k + p - 1)$ dimensional state vector be defined by $x_n = (t_n, \cdots, t_{n-k+1}, s_n, s_{n-1}, \cdots, s_{n-p+2})^T$, and define a two-dimensional noise vector as $v_n = (v_{n1}, v_{n2})^T$, and the matrices F, G and H by

$$
F = \begin{bmatrix} F_1 & O \\ O & F_2 \end{bmatrix}, \quad
G = \begin{bmatrix} G_1 & O \\ O & G_2 \end{bmatrix}, \quad
H = [H_1 \ H_2], \quad (12.18)
$$

then the state-space representation of the seasonal adjustment model is obtained as

$$
\begin{aligned}
x_n &= F x_{n-1} + G v_n \\
y_n &= H x_n + w_n.
\end{aligned} \qquad (12.19)
$$

In (12.18), F_1, G_1 and H_1 are the matrices and vectors used for the state-space representation of the trend component model, and similarly, F_2, G_2 and H_2 are the matrices and vectors used for the representation of the seasonal component model. For instance, if $k = 2$, $\ell = 1$ and $p = 4$, then F, G and H are defined by

$$
F = \begin{bmatrix}
2 & -1 & & & \\
1 & 0 & & & \\
& & -1 & -1 & -1 \\
& & 1 & & \\
& & & 1 &
\end{bmatrix}, \quad
G = \begin{bmatrix}
1 & 0 \\
0 & 0 \\
0 & 1 \\
0 & 0 \\
0 & 0
\end{bmatrix}
$$

$$H = \begin{bmatrix} 1 & 0 & 1 & 0 & 0 \end{bmatrix}. \qquad (12.20)$$

Example (Seasonal adjustment of WHARD data)

R function `season` of TSSS package fits the seasonal adjustment model and obtains trend and seasonal components. Various standard seasonal adjustment model can be obtained by specifying the following optional arguments:

`trend.order:`	trend order (0, 1, 2 or 3).
`seasonal.order:`	seasonal order (0, 1 or 2).
`period:`	If the tsp attribute of y is NULL, period length of one season is required.
`tau2.ini:`	initial estimate of variance of the system noise τ^2, not equal to 1.
`filter:`	a numerical vector of the form c(x1,x2) which gives start and end position of filtering.
`predict:`	the end position of prediction (>= x2).
`log:`	logical. If TRUE, the data y is log-transformed.
`minmax:`	lower and upper limits of observations.
`plot:`	logical. If TRUE (default), 'trend' and 'seasonal' are plotted.

Note that this seasonal adjustment method can handle not only the monthly data, but also verious period length. For example, set `period=4` for quarterly data, = 12 for monthly data, =5 or 7 for weekly data and =24 for hourly data.

The function `season` yields the following outputs:

`tau2:`	variance of the system noise.
`sigma2:`	variance of the observational noise.
`llkhood:`	log-likelihood of the model.
`aic:`	AIC of the model.
`trend:`	trend component (for `trend.order` > 0).
`seasonal:`	seasonal component (for `seasonal.order` > 0).
`noise:`	noise component.
`cov:`	covariance matrix of smoother.

```
> # Wholesale hardware data
> data(WHARD)
> season(WHARD, trend.order =2, seasonal.order =1, log =TRUE)
```

Figure 12.1 shows the estimates of the trend, the seasonal and the noise components of the logarithm of WHARD data using the standard seasonal adjustment model. The maximum likelihood estimates of the

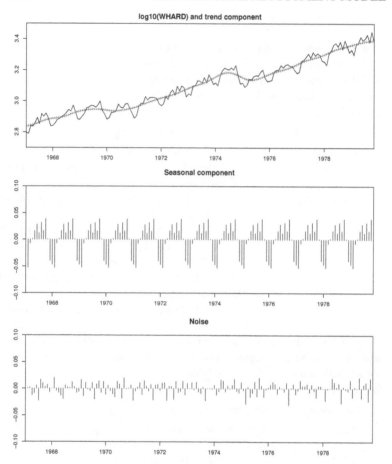

Figure 12.1 *Seasonal adjustment of WHARD data by a standard seasonal adjustment model.*

parameters are $\hat{\tau}_1^2 = 0.0248$, $\hat{\tau}_2^2 = 0.11 \times 10^{-7}$, $\hat{\sigma}^2 = 0.156 \times 10^{-3}$, and AIC $= -728.50$. Plot (b) in Figure 12.1 shows the estimated trend component $t_{n|N}$ and ± 1 standard error interval. A very smooth upward trend is evident, except for the years 1974 and 1975 where a rapid decrease in the trend is detected. On the other hand, the estimated seasonal component is very stable over the whole interval.

The function season can also perform long-term prediction based on the estimated seasonal adjustment model. For that purpose, use the

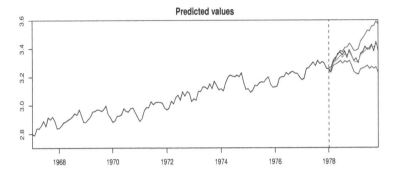

Figure 12.2: *Long-term prediction using a seasonal adjustment model.*

argument `filter` to specify the data period to be used for the estimation of the parameters. In the example below, `filter=c(1,132)` indicates that the seasonal adjustment model is estimated from the data $y_1, \ldots, y_{132}$ and the rest of the data $y_{133}, \ldots, y_N$ are predicted based on the estimated model.

```
> # Wholesale hardware data
> data(WHARD)
> season(WHARD, trend.order = 2, seasonal.order = 1, filter =
c(1,132), log = TRUE)
```

Using the seasonal adjustment model, we can predict time series with seasonal variation. Figure 12.2 shows the long-term prediction for two years, i.e. 24 observations using the initial 132 observations by the method shown in Chapter 9. In the figure, the mean of the predictive distribution $y_{132+j|132}$ and the ± 1 standard error interval, i.e., $\pm \sqrt{d_{132+j|132}}$ for $j = 1, \ldots, 24$ are shown. The actual observations are also shown. In this case, very good long-term prediction can be attained, because the estimated trend and seasonal components are steady. However, it should be noted that ± 1 standard error interval is very wide indicating that the long-term prediction by this model is not so reliable.

12.3 Decomposition Including a Stationary AR Component

In this section, we consider an extension of the standard seasonal adjust-ment method (Kitagawa and Gersch (1984)). In the standard seasonal adjustment method, the time series is decomposed into three components, i.e. the trend component, the seasonal component and the observation noise. These components are assumed to follow the models given in (12.15) and (12.16) and the observation noise is assumed to be a white noise. Therefore, if a significant deviation from that assumption is present, then the decomposition obtained by the standard seasonal adjustment method might become inappropriate.

```
> data(BLSALLFOOD)
> season(BLSALLFOOD, trend.order = 2, seasonal.order = 1)
```

Figure 12.3 shows the decomposition of the BLSALLFOOD data of Figure 1.1(d) that was obtained by the standard seasonal adjustment method for the model with $k = 2$, $\ell = 1$ and $p = 12$. In this case, different from the case shown in the previous section, the estimated trend shows a wiggle, particularly in the latter part of the data.

```
> data(BLSALLFOOD)
> season(BLSALLFOOD, trend.order = 2, seasonal.order = 1,
filter = c(1,132))
```

Let us consider the problems when the above-mentioned wiggly trend is obtained. Similarly to the Figure 12.2, Figure 12.4 shows the long-term prediction with a two-year prediction horizon (24 observations) of the BLSALLFOOD data based on the former 132 observations. In this case, apparently the predicted mean $y_{132+j|132}$ provides a reasonable prediction of the actual time series y_{132+j}. However it is evident that prediction by this model is not reliable, because an explosive increase of the width of the confidence interval is observed.

Figure 12.5 shows the overlay of 13 long-term predictions that are obtained by changing the starting point to be $n = 126, \ldots, 138$. The long-term predictions starting at and before $n = 130$ have significant downward bias. On the other hand, the long-term predictions starting at and after $n = 135$ have significant upward bias. This is the reason that the explosive increase in the width of the confidence interval has occurred when the lead time has increased in the long-term predictions. The stochastic trend component model in the seasonal adjustment model can flexibly express a complex trend component. But in the long-term prediction with this model, the predicted mean $t_{n+j|n}$ is simply obtained by using the difference equation

$$\Delta^k t_{n+j|n} = 0. \tag{12.21}$$

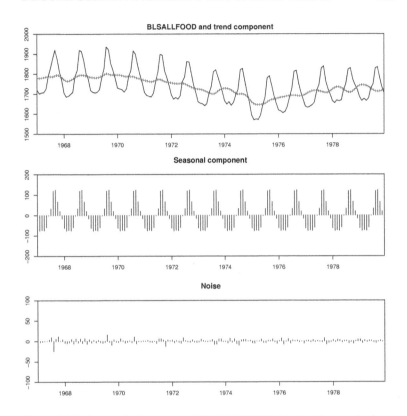

Figure 12.3 *Seasonal adjustment of BLSALLFOOD data by the standard seasonal adjustment model.*

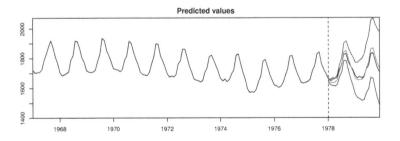

Figure 12.4 *Long-term prediction of BLSALLFOOD data (prediction starting point: n = 132).*

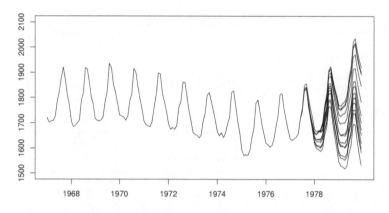

Figure 12.5 *Long-term prediction of the BLSALLFOOD data with floating start-ing points. Prediction starting points are* $n = 126,\dots,138$.

Therefore, whether the predicted values go up or down is decided by the starting point of the trend. From these results, we see that if the esti-mated trend is wiggly, it is not appropriate to use the standard seasonal adjustment model for long-term prediction.

In predicting one year or two years ahead, it is possible to obtain a better prediction with a smoother curve by using a smaller value than the maximum likelihood estimate for the system noise variance of the trend component, τ_1^2. However, this method suffers from the following prob-lems; it is difficult to reasonably determine τ_1^2 and, moreover, prediction with a small lead time such as one-step-ahead prediction becomes sig-nificantly worse than that obtained by the maximum likelihood model.

To achieve good prediction for both short and long lead times, we consider an extension of the standard seasonal adjustment model by including a new component p_n as

$$y_n = t_n + s_n + p_n + w_n. \tag{12.22}$$

Here p_n is a *stationary AR component model* that is assumed to follow an AR model

$$p_n = \sum_{i=1}^{m_3} a_i p_{n-i} + v_{n3}, \tag{12.23}$$

where v_{n3} is assumed to be a Gaussian white noise with mean 0 and vari-ance τ_3^2. This model expresses a short-term variation such as the cycle of an economic time series, not a long-term tendency like the trend component. In the model (12.22), the trend component obtained by the standard seasonal adjustment model is further decomposed into the smoother trend component t_n and the short-term variation p_n. As shown in Chapter 9, the state-space representation of the AR model (12.22) is obtained by defining the state vector $x_n = (p_n, p_{n-1}, \ldots, p_{n-m_3+1})^T$ and

$$F_3 = \begin{bmatrix} a_1 & a_2 & \cdots & a_{m_3} \\ 1 & & & \\ & & \ddots & \\ & & & 1 \end{bmatrix}, \quad G_3 = \begin{bmatrix} 1 \\ 0 \\ \vdots \\ 0 \end{bmatrix}, \quad (12.24)$$

$$H_3 = [\, 1 \quad 0 \quad \cdots \quad 0 \,], \qquad Q_3 = \tau_3^2.$$

Therefore, using the state-space representation for the composite model shown in Chapter 9, the state-space model for the decomposition of (12.22) is obtained by defining

$$F = \begin{bmatrix} F_1 & & \\ & F_2 & \\ & & F_3 \end{bmatrix}, \quad G = \begin{bmatrix} G_1 & 0 & 0 \\ 0 & G_2 & 0 \\ 0 & 0 & G_3 \end{bmatrix},$$

$$H = [\, H_1 \quad H_2 \quad H_3 \,], \quad Q = \begin{bmatrix} \tau_1^2 & 0 & 0 \\ 0 & \tau_2^2 & 0 \\ 0 & 0 & \tau_3^2 \end{bmatrix}. \quad (12.25)$$

Example (Seasonal adjustment with stationary AR component)

To fit a seasonal adjustment model including an AR component by the function season, it is only necessary to specify the AR order by the ar.order parameter as follows.

```
> data(BLSALLFOOD)
> season(BLSALLFOOD, trend.order = 2, seasonal.order = 1,
ar.order = 2)
```

Figure 12.6 shows the decomposition of BLSALLFOOD data into trend, seasonal, stationary AR and observation noise components by this model. The estimated trend expresses a very smooth curve similar to Figure 12.1. On the other hand, a short-term variation is detected as the stationary AR component and the (trend component) + (stationary AR component) resemble the trend component of Figure 12.3. The AIC for this model is 1336.54, which is significantly smaller than that of the standard seasonal adjustment model, 1369.30.

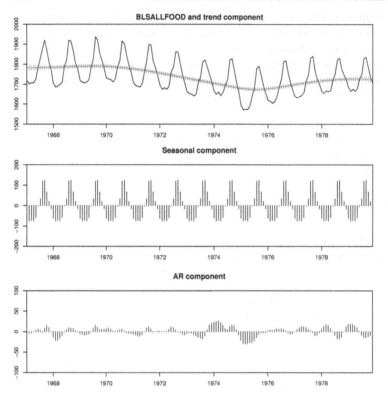

Figure 12.6 *Seasonal adjustment of BLSALLFOOD data by the model including stationary AR components.*

```
> data(BLSALLFOOD)
> season(BLSALLFOOD, trend.order = 2, seasonal.order = 1,
ar.order = 2, filter = c(1,132))
```

Figure 12.7 shows the long-term prediction with this model starting from $n = 132$. Comparing this with Figure 12.7, it can be seen that a good prediction was achieved for both short and long lead times with this modeling.

12.4   Decomposition Including a Trading-Day Effect

Monthly economic time series, such as the amount of sales at a department store, may strongly depend on the number of the days of the week

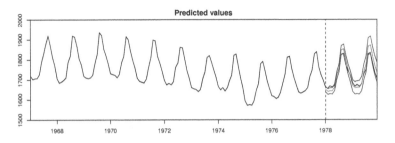

Figure 12.7 *Long-term prediction of the BLSALLFOOD data by the model including stationary AR components.*

in each month, because there are marked differences in sales depending on the days of the week. For example, a department store would have more customers on Saturdays and Sundays or may regularly close on a specific day of the week. A similar phenomenon can often be seen in environmental data, for example, the daily amounts of NO_x and CO_2.

A *trading-day adjustment* has been developed to remove such trading-day effects that depend on the number of the days of the week (Cleveland and Devlin (1980), Hillmer (1982)). To develop a model-based method for trading-day adjustment, we need a proper model for the trading-day effect component (Akaike and Ishiguro (1983), Kitagawa and Gersch (1984, 1996)). Hereafter, the number of Sundays to Saturdays in the n-th month, y_n, are denoted as $d_{n1}^*, \cdots, d_{n7}^*$. Note that each d_{ni}^* takes a value 4 or 5 for monthly data.

Then the effect of the trading-day is expressed as

$$td_n = \sum_{i=1}^{7} \beta_{ni} d_{ni}^*. \tag{12.26}$$

The coefficient β_{ni} expresses the effect of the number of the i-th day of the week on the value of y_n. Here, to guarantee the uniqueness of the decomposition of the time series, we impose the restriction that the sum of all coefficients amounts to 0, that is,

$$\beta_{n1} + \cdots + \beta_{n7} = 0. \tag{12.27}$$

Then, since the last coefficient is defined by $\beta_{n7} = -(\beta_{n1} + \cdots + \beta_{n6})$, the trading-day effect can be expressed by using $\beta_{n1}, \ldots, \beta_{n6}$ as

$$td_n = \sum_{i=1}^{6} \beta_{ni}(d_{ni}^* - d_{n7}^*)$$

$$\equiv \sum_{i=1}^{6} \beta_{ni} d_{ni}, \tag{12.28}$$

where $d_{ni} \equiv d_{ni}^* - d_{n7}^*$ is the difference in the number of the i-th day of the week and the number of Saturdays in the n-th month, y_n.

On the assumption that these coefficients gradually change, following the first order trend model

$$\Delta\beta_{ni} = v_{n4}^{(i)}, \qquad v_{n4}^{(i)} \sim N(0, \tau_4^2), \qquad i = 1, \ldots, 6, \tag{12.29}$$

the state-space representation of the trading-day effect model is obtained by

$$F_{n4} = G_{n4} = I_6, \qquad H_{n4} = [d_{n1}, \ldots, d_{n6}]$$

$$x_{n4} = \begin{bmatrix} \beta_{n1} \\ \vdots \\ \beta_{n6} \end{bmatrix}, \qquad Q = \begin{bmatrix} \tau_4^2 & & \\ & \ddots & \\ & & \tau_4^2 \end{bmatrix}. \tag{12.30}$$

In this representation, H_{n4} becomes a time-dependent vector. For simplicity, the variance covariance matrix Q is assumed to be a diagonal matrix with equal diagonal elements. In actual analysis, however, we often assume that the coefficients are time-invariant, i.e.,

$$\beta_{ni} \equiv \beta_i, \tag{12.31}$$

which can be implemented by putting either $\tau_4^2 = 0$ or $G = 0$ in the state-space representation of the model.

The state-space model for the decomposition of the time series

$$y_n = t_n + s_n + p_n + td_n + w_n \tag{12.32}$$

is obtained by using

$$x_n = \begin{bmatrix} x_{n1} \\ x_{n2} \\ x_{n3} \\ x_{n4} \end{bmatrix}, F = \begin{bmatrix} F_1 & & & \\ & F_2 & & \\ & & F_3 & \\ & & & F_4 \end{bmatrix}, G = \begin{bmatrix} G_1 & 0 & 0 \\ 0 & G_2 & 0 \\ 0 & 0 & G_3 \\ 0 & 0 & 0 \end{bmatrix},$$

$$H = [\ H_1 \quad H_2 \quad H_3 \quad H_{n4}\ ], \quad Q = \begin{bmatrix} \tau_1^2 & 0 & 0 \\ 0 & \tau_2^2 & 0 \\ 0 & 0 & \tau_3^2 \end{bmatrix}. \tag{12.33}$$

Example (Seasonal adjustment with trading-day effect)

To perform the trading-day adjustment with the R function `season`, we set `trade=TRUE` as follows:

```
> season(WHARD, trend.order = 2, seasonal.order = 1, ar.order =
0, trade = TRUE, log = TRUE)
```

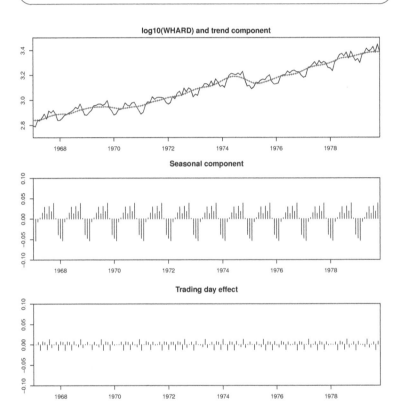

Figure 12.8: *Trading-day adjustment for WHARD data.*

Figure 12.8 shows the logarithm of the WHARD data, the decomposition into the trend, the seasonal, the trading-day effect and the noise components by the seasonal adjustment model with trading-day effect. The estimated trend and seasonal components are similar to the decomposition by the standard seasonal adjustment model shown in Figure 12.1. Although the extracted trading-day effect component is apparently minuscule, it can be seen that the plot of seasonal component plus trading-day effect component reproduces the details of the observed time series. The AIC value for this model is -778.18 which is significantly smaller than that for the standard seasonal adjustment model, -728.50. Once the time-invariant coefficients of trading-day effects β_i are estimated, this can contribute to a significant increase in the accuracy of prediction, because $d_{n1}, \cdots, d_{n6}$ are known even for the future.

The function season can perform the decomposition into trend, seasonal and trading-day effect components. In the R code below, they are given in the arguments z$trend, z$season and z$day.effect, respectively. Figure 12.9 shows the seasonal component, trading-day effect, seasonal component plus trading-day effect and log-transformed WHARD data minus trend component, respectively. It can be seen that, although the trading-day effect is minuscule, the seasonal component plus trading-day effect is quite similar to the data minus trend component. This suggests that a considerable part of the noise component of the standard seasonal adjustment model can be expressed as the trading-day effect.

```
> data(WHARD)
> z <- season(WHARD, trend.order = 2, seasonal.order = 1,
ar.order = 0, trade = TRUE, log = TRUE)
> a1 <- z$seasonal + z$day.effect
> a2 <- log10( as.numeric(WHARD) )
> a3 <- a2 - z$trend
> par(mar=c(2,2,3,1)+0.1)
> par( mfrow=c(4,1) )
> plot( z$seasonal,type="l", ylim=c(-0.1,0.1), main="seasonal
component" )
> plot( z$day.effect,type="l", ylim=c(-0.1,0.1),
main="trading-day effect" )
> plot( a1,type="l", ylim=c(-0.1,0.1), main="seasonal +
trading-day effect" )
> plot( a3,type="l", ylim=c(-0.1,0.1), main="data - trend" )
```

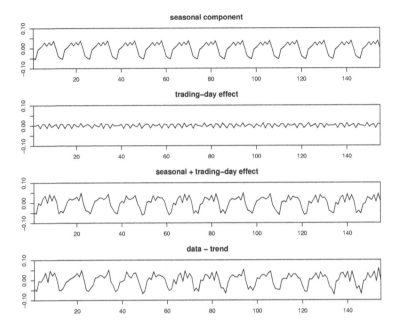

Figure 12.9 *Trading-day effect. From top to the bottom, seasonal component, trading-day effect, seasonal component plus trading-day effect and data minus trend.*

Problems

1. Explain what to keep in mind when decomposing a time series using a seasonal adjustment model with a stationary AR component.

2. Consider a different seasonal component model from that given in 12.1.

3. In trading day adjustments, consider a model that takes into account the following components,

   (1) The effects of the weekend (Saturday and Sunday) and weekdays are different.

(2) There are three different effects, i.e., Saturday, Sunday and the weekdays.

4. Consider any other possible effects related to seasonal adjustments that are not considered in this book.

5. Try seasonal adjustment of an actual time series.

Chapter 13

Time-Varying Coefficient AR Model

There are two types of nonstationary time series, one with a drifting mean value and the other with a varying structure around the mean value function. For the latter type of nonstationary time series, the variance, the autocovariance function, or the power spectrum of the time series change over time.

In this chapter, two methods are presented for the analysis of such nonstationary time series. One is an estimation method for time-varying variance, and the other is a method for modeling the time-varying coefficient AR model. The estimation of the stochastic volatility in financial time series analysis can be considered as a typical example of the estimation of the time-varying variance.

Early treatment of AR model with time-varying coefficients are reported in Whittle (1965), Subba Rao (1970) and Bohlin (1976). A state-space modeling for the time-varying AR model was introduced in Kitagawa (1983) and Kitagawa and Gersch (1985,1996). Extension of the state-space modeling to multivariate time series was shown in Jiang and Kitagawa (1993).

13.1 Time-Varying Variance Model

The time series y_n, $n = 1, \ldots, N$, is assumed to be a Gaussian white noise with mean 0 and time-varying variance σ_n^2. Assuming that $\sigma_{2m-1}^2 = \sigma_{2m}^2$, if we define a transformed series $s_1, \ldots, s_{N/2}$ by

$$s_m = y_{2m-1}^2 + y_{2m}^2, \qquad (13.1)$$

s_m is distributed as a χ^2-distribution with 2 degrees of freedom which is known to be an exponential distribution. Therefore, the probability density function of s_m is given by

$$f(s) = \frac{1}{2\sigma^2} e^{-s/2\sigma^2}. \qquad (13.2)$$

213

The probability density function of the random variable z_m defined by the transformation

$$z_m = \log\left(\frac{s_m}{2}\right),$$ (13.3)

is expressed as

$$g(z) = \frac{1}{\sigma^2}\exp\left\{z - \frac{e^z}{\sigma^2}\right\} = \exp\left\{(z - \log\sigma^2) - e^{(z - \log\sigma^2)}\right\}.$$ (13.4)

This means that the transformed series z_m can be written as

$$z_m = \log\sigma_m^2 + w_m,$$ (13.5)

where w_m follows a double exponential distribution with probability density function

$$h(w) = \exp\left\{w - e^w\right\}.$$ (13.6)

Therefore, assuming that the logarithm of the time-varying variance, $\log\sigma_m^2$, follows a k-th order trend component model, the transformed series z_m can be expressed by a state-space model,

$$\begin{aligned} \Delta^k t_m &= v_m \\ z_m &= t_m + w_m, \end{aligned}$$ (13.7)

and the logarithm of the variance of the original time series y_n is estimated by obtaining the trend of the transformed series z_m.

It should be noted that the distribution of the noise w_m is not Gaussian. However, since the mean and the variance of the double exponential distribution are given by $-\zeta = 0.57722$ (Euler constant) and $\pi^2/6$, respectively, we can approximate it by a Gaussian distribution as follows:

$$w_m \sim N\left(-\zeta, \frac{\pi^2}{6}\right).$$ (13.8)

Wahba (1980) used this property in smoothing the log periodogram with a cross-validated smoothing spline. With this approximation, we can estimate the trend t_m with the Kalman filter. Then $t_{m|M} + \gamma$, $m = 1, \ldots, M$ with $M = N/2$ yields a smoothed estimate of $\log\sigma_m^2$. Later in Chapter 14, an exact method of estimating the logarithm of the variance will be given, using the exact probability distribution (13.6) and a non-Gaussian filter and a smoother.

So far, the transformed series defined as the mean of two consecutive squared time series shown in equation (13.1) has been used in estimating the variance of the time series. This is just to make the noise distribution $g(z)$ closer to a Gaussian distribution, and it is not essential in estimating the variance of the time series. In fact, if we use a non-Gaussian filter for trend estimation, we can use the transformed series $s_n = y_n^2$, $n = 1,\ldots,N$, and with this transformation, it is not necessary to assume that $\sigma_{2m-1}^2 = \sigma_{2m}^2$.

Example (Time-varying variance of a seismic data)

R function `tvvar` of the package TSSS estimates time-varying variance. The argument `trend.order` specifies the order of the trend component model, `tau2.ini` and `delta`, respectively, specify the initial estimate of the variance of the system noise τ^2 and the search width for finding the maximum likelihood estimate. If they are not given explicitly, i.e. if `tau2.ini` = NULL, the most suitable value is chosen within the candidates $\tau^2 = 2^{-k}$, $k = 1,2,\ldots$.

Then the function `tvvar` returns the following outputs:

tvv:	time-varying variance
nordata:	normalized data
sm:	transformed data
trend:	trend of the transformed data sm
noise:	residuals
tau2:	varaince of the system noise
sigma2:	variance of the observation noise
llkhood:	log-likelihood of the model
aic:	AIC of the model
tsname:	the name of the time series y_n

```
> data(MYE1F)
> tvvar(MYE1F, trend.order = 2, tau2.ini = 6.6e-06, delta =
1.0e-06)
```

The function `tvvar` was applied to the MYE1F seismic data shown in Figure 1.1 using trend order 2 and the initial variance $\tau^2 = 0.66 \times 10^{-5}$. Figure 13.1 shows the transformed data s_m, the estimates of the logarithm of the variance $\log \hat{\sigma}_m^2$, the residual $s_m - \log \hat{\sigma}_m^2$ and the normalized time series obtained by $\tilde{y}_n = \hat{\sigma}_{n/2}^{-1} y_n$. By this normalization, we can obtain a time series with a variance roughly equal to 1.

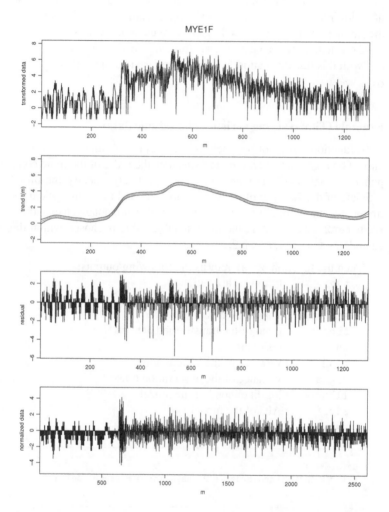

Figure 13.1 *Estimation of time-varying variance and standardized time series. From top to bottom: transformed data, estimated time-varying log-variance and the normalized time series.*

13.2 Time-Varying Coefficient AR Model

The characteristics of stationary time series can be expressed by an auto-covariance function or a power spectrum. Therefore, for nonstationary time series with a time-varying stochastic structure, it is natural to consider that its autocovariance function and power spectrum change over time. For a stationary time series, its autocovariance function and power spectrum are characterized by properly specifying the orders and coefficients of an AR model or ARMA model. Therefore, for a nonstationary time series with a time-varying stochastic structure, it is natural to consider that these coefficients change with time.

In this section, an autoregressive model with time-varying coefficients for the nonstationary time series y_n is modeled as

$$y_n = \sum_{j=1}^{m} a_{nj} y_{n-j} + w_n, \qquad (13.9)$$

where w_n is a Gaussian white noise with mean 0 and variance σ^2 (Kozin and Nakajima (1980), Kitagawa (1983)). This model is called the *time-varying coefficients AR model* of order m and a_{nj} is called the *time-varying AR coefficient* with time lag j at time n. Given the time series $y_1,\ldots,y_N$, this time-varying coefficients AR model contains at least mN unknown coefficients. The difficulty with this type of model is, therefore, that we cannot obtain meaningful estimates by applying the ordinary maximum likelihood method or the least squares method to the model (13.9).

To circumvent this difficulty, similar to the cases of the trend model and the seasonal adjustment model, we use a stochastic trend component model to represent time-varying parameters of the AR model. In the case of the trend model, we considered the trend component t_n as an unknown parameter and introduced the model for its time change. Similarly, in the case of a time-varying coefficient AR model, since the AR coefficient a_{nj} changes over time n, we consider a constraint model

$$\Delta^k a_{nj} = v_{nj}, \qquad j = 1,\ldots,m, \qquad (13.10)$$

where Δ is the difference operator with respect to time n defined as $\Delta a_{nj} = a_{nj} - a_{n-1,j}$ (Kitagawa (1983), Kitagawa and Gersch (1985, 1996)).

In (13.10), k is assumed to be 1 or 2. The vector $v_n = (v_{n1},\ldots,v_{nm})^T$ is an m-dimensional Gaussian white noise with mean vector 0 and variance covariance matrix Q. Since v_{ni} and v_{nj} are usually assumed to be independent for $i \neq j$, Q becomes a diagonal matrix with diagonal elements $\tau_{11}^2, \ldots, \tau_{mm}^2$, thus it can be expressed as $Q = \mathrm{diag}\{\tau_{11}^2, \ldots, \tau_{mm}^2\}$. It is assumed further in this section that $\tau_{11}^2 = \cdots = \tau_{mm}^2 = \tau^2$. The rationale for this assumption will be discussed later in Section 13.4.

Next, to estimate the AR coefficients of the time-varying AR model (13.9) associated with the component model (13.10), we develop a corresponding state-space representation. For $k = 1$ and $k = 2$, the state vectors are defined by $x_{nj} = a_{nj}$ and $x_{nj} = (a_{nj}, a_{n-1,j})^T$, respectively. Then the time-varying coefficient AR model in equation (13.10) can be expressed as

$$x_{nj} = F^{(k)} x_{n-1,j} + G^{(k)} v_{nj}, \tag{13.11}$$

where $F^{(1)}$, $F^{(2)}$, $G^{(1)}$ and $G^{(2)}$ are defined as

$$F^{(1)} = G^{(1)} = 1,$$

$$F^{(2)} = \begin{bmatrix} 2 & -1 \\ 1 & 0 \end{bmatrix}, \quad G^{(2)} = \begin{bmatrix} 1 \\ 0 \end{bmatrix}. \tag{13.12}$$

Here, assuming the j-th term of equation (13.9) to be the j-th component of this model, it can be expressed as

$$a_{nj} y_{n-j} = H_n^{(kj)} x_{nj}, \tag{13.13}$$

where $H_n^{(1j)} = y_{n-j}$ and $H_n^{(2j)} = (y_{n-j}, 0)$. Then the j-th component of the time-varying coefficient AR model is given by a state-space model with $F^{(k)}$, $G^{(k)}$ and $H_n^{(kj)}$ as follows:

$$\begin{aligned} x_{nj} &= F^{(k)} x_{n-1,j} + G^{(k)} v_{nj} \\ a_{nj} y_{n-j} &= H_n^{(kj)} x_{nj}. \end{aligned} \tag{13.14}$$

Moreover, noting that $H^{(1)} = 1$ and $H^{(2)} = (1,0)$, $H_n^{(kj)}$ is given by

$$H_n^{(kj)} = y_{n-j} H^{(k)}. \tag{13.15}$$

Using the above component models, a state-space representation of the time-varying coefficients AR model is obtained as

$$\begin{aligned} x_n &= F x_{n-1} + G v_n \\ y_n &= H_n x_n + w_n, \end{aligned} \tag{13.16}$$

where the $km \times km$ matrix F, the $km \times m$ matrix G and the km-dimensional vectors H_n and x_n are defined by

$$F = \begin{bmatrix} F^{(k)} & & \\ & \ddots & \\ & & F^{(k)} \end{bmatrix} = I_m \otimes F^{(k)}$$

$$G = \begin{bmatrix} G^{(k)} & & \\ & \ddots & \\ & & G^{(k)} \end{bmatrix} = I_m \otimes G^{(k)}$$

$$H_n = [H_n^{(k1)}, \ldots, H_n^{(km)}] = (y_{n-1}, \ldots, y_{n-m}) \otimes H^{(k)}$$

$$x_n = \begin{cases} (a_{n1}, \ldots, a_{nm})^T, & \text{for } k = 1 \\ (a_{n1}, a_{n-1,1}, \ldots, a_{nm}, a_{n-1,m})^T, & \text{for } k = 2 \end{cases}$$

$$Q = \begin{bmatrix} \tau^2 & & \\ & \ddots & \\ & & \tau^2 \end{bmatrix}, \quad R = \sigma^2. \quad (13.17)$$

Here, I_m is the $m \times m$ unit matrix and $\otimes$ denotes the Kronecker product of the matrices A and B, i.e., for the $k \times \ell$ matrix $A = (a_{ij})$ and the $m \times n$ matrix $B = (b_{kl})$, $A \otimes B$ is the $km \times \ell n$ matrix defined by

$$A \otimes B = \begin{bmatrix} a_{11}B & \cdots & a_{1n}B \\ \vdots & \ddots & \vdots \\ a_{m1}B & \cdots & a_{mn}B \end{bmatrix}.$$

Hence the time-varying coefficients AR model (13.9) and the component model for the time-varying coefficients are expressible in the form of a state-space model.

For instance, for $m = 2$ and $k = 2$, the state-space model is defined as

$$\begin{bmatrix} a_{n1} \\ a_{n-1,1} \\ a_{n2} \\ a_{n-1,2} \end{bmatrix} = \begin{bmatrix} 2 & -1 & 0 & 0 \\ 1 & 0 & 0 & 0 \\ 0 & 0 & 2 & -1 \\ 0 & 0 & 1 & 0 \end{bmatrix} \begin{bmatrix} a_{n-1,1} \\ a_{n-2,1} \\ a_{n-1,2} \\ a_{n-2,2} \end{bmatrix} + \begin{bmatrix} 1 & 0 \\ 0 & 0 \\ 0 & 1 \\ 0 & 0 \end{bmatrix} \begin{bmatrix} v_{n1} \\ v_{n2} \end{bmatrix}$$

$$y_n = (y_{n-1}, 0, y_{n-2}, 0) \begin{bmatrix} a_{n1} \\ a_{n-1,1} \\ a_{n2} \\ a_{n-1,2} \end{bmatrix} + w_n \quad (13.18)$$

$$\begin{bmatrix} v_{n,1} \\ v_{n,2} \end{bmatrix} \sim N\left(\begin{bmatrix} 0 \\ 0 \end{bmatrix}, \begin{bmatrix} \tau^2 & 0 \\ 0 & \tau^2 \end{bmatrix} \right), \qquad w_n \sim N(0, \sigma^2).$$

The above state-space model contains several unknown parameters. For these parameters, since the log-likelihood function can be computed by the method presented in Chapter 9, the variance σ^2 of the observation noise w_n and the variance τ^2 of the system noise v_n can be estimated by the maximum likelihood method. The AR order m and the order k of the smoothness constraint model (13.10) can be determined by minimizing the information criterion AIC. Given the orders m and k and the estimated variances $\hat{\sigma}^2$ and $\hat{\tau}^2$, the smoothed estimate of the state vector $x_{n|N}$ is obtained by the fixed interval smoothing algorithm. Then the $((j-1)k+1)$-th element of the estimated state gives the smoothed estimate of the time-varying AR coefficient $\hat{a}_{n,j|N}$.

Storage of size $mk \times mk \times N$ is necessary to obtain the smoothed estimates of these time-varying AR coefficients simultaneously. If either the AR order m or the series length N is large, the necessary memory size may exceed the memory of the computer and would make the computation impossible. Making an assumption that the AR coefficients change only once every r time-steps for some integer $r > 1$ may be the simplest method of mitigating such memory problems of the computing facilities. For example, if $r = 20$, the necessary memory size is obviously reduced by a factor of 20. If the AR coefficients change slowly and gradually, such an approximation has only a slight effect on the estimated spectrum. To execute the Kalman filter and smoother for $r > 1$, we repeat the filtering step r times at each step of the one-step-ahead prediction.

Example (Time-varying coefficient AR models for seismic data)

The R function tvar of the package TSSS estimates the time-varying coefficient AR model. The inputs to the function are

y:	a univariate time series.
trend.order:	trend order (1 or 2).
ar.order:	AR order.
span:	local stationary span r.
outlier:	position of outliers.
tau2.ini:	initial estimate of the system noise τ^2. If tau2.ini=NULL the most suitable value is chosen when $\tau^2 = 2^{-k}, k = 1, 2, \dots$.
delta:	search width.
plot:	logical. If TRUE (default), PARCOR is plotted.

and the outputs from this function are:

`arcoef`:	time-varying AR coefficients.
`sigma2`:	variance of the observation noise σ^2.
`tau2`:	variance of the system noise τ^2.
`llkhood`:	log-likelihood of the model.
`aic`:	AIC of the model.
`parcor`:	time-varying partial autocorrelation coefficients

```
> data(MYE1F)
> # k=1
> z <- tvar(MYE1F, trend.order = 1, ar.order = 8,
span = 40, tau2.ini = 1.0e-03, delta = 1.0e-04)
> # k=2
> z <- tvar(MYE1F, trend.order = 2, ar.order = 8,
span = 40, tau2.ini = 6.6e-06, delta = 1.0e-06)
```

Table 13.1 summarizes the AIC's of time-varying coefficient AR models fitted to the normalized seismic data shown in Figure 13.1, with various orders obtained by putting $r = 40$.

Since the observation noise variance σ^2 is assumed to be a constant in the TVCAR modeling, if the variance of the time series significantly changes as can be observed in Figure 1.1(f), it is better to fit the model to a transformed time series. For instance, Figure 13.1 shows that the variance of the time series is approximately homoscedastic. For this data, the AIC was minimized at $m = 8$ for $k = 1$ and $m = 4$ for $k = 2$.

Table 13.1 *AIC's of time-varying coefficients AR models fitted to normalized seismic data.*

m	$k = 1$	$k = 2$	m	$k = 1$	$k = 2$
1	6492.5	6520.4	6	4831.9	4873.8
2	5527.7	5643.2	7	4821.6	4878.7
3	5070.0	5134.5	8	4805.1	4866.9
4	4820.0	4853.9	9	4813.4	4884.9
5	4846.0	4886.0	10	4827.1	4911.9

Figure 13.2 shows the time-varying coefficients of the TVCAR models with the AIC best orders m for $k = 1$ and $k = 2$. Note that the figure shows the time-varying PARCOR's b_{ni} for $i = 1, 2, 3, 4$, instead of the AR coefficients. The plots on the left-hand side are for the case of $k = 1$ and on the right-hand side are for the case of $k = 2$. From the figure, we see that the estimated coefficients vary with time corresponding to the

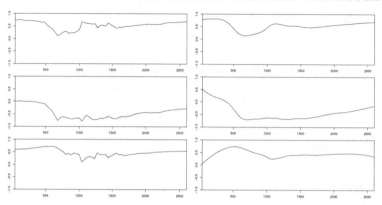

Figure 13.2 *Estimated time-varying PARCOR for the normalized seismic data.*
Only the first four PARCOR's are shown. Left plots: $k = 1$, right plots: $k = 2$.
Horizontal axis: time point, vertical axis: value of PARCOR.

arrivals of the P-wave and the S-wave at $n = 630$ and $n = 1026$, respectively, which were estimated by using the locally stationary AR model.

In Figure 13.2, the estimates for $k = 2$ are very smooth. However,
as is apparent in the figure, the estimates with $k = 2$ cannot reveal the
abrupt changes of the time series that correspond to the arrival of the P-
wave and the S-wave. Compared with this, the estimates obtained from
the TVCAR model with $k = 1$ have larger fluctuations. According to the
AIC, the TVCAR model with $k = 1$ is considered to be a better model
than that with $k = 2$. In Section 13.5, we shall consider a method of
obtaining smooth estimates that also has the capability to adapt to abrupt
structural changes.

13.3 Estimation of the Time-Varying Spectrum

For a stationary AR model, the power spectrum is given by

$$p(f) = \frac{\sigma^2}{\left|1 - \sum_{j=1}^{m} a_j e^{-2\pi i j f}\right|^2}, \qquad -\frac{1}{2} \le f \le \frac{1}{2}. \qquad (13.19)$$

Therefore, for the time-varying coefficient AR model (13.9) with time-
varying AR coefficients a_{nj}, the instantaneous spectrum at time n can be

defined by

$$p_n(f) = \frac{\sigma^2}{\left|1 - \sum_{j=1}^{m} a_{nj} e^{-2\pi i j f}\right|^2}, \qquad -\frac{1}{2} \leq f \leq \frac{1}{2}. \qquad (13.20)$$

Thus, using the time-varying AR coefficients a_{nj} introduced in the previous section, we can estimate the time-varying power spectrum as a function of time, which is called the *time-varying spectrum*.

Example (Time-varying spectrum of a seismic data)

The R function tvspec of the package TSSS computes time-varying spectrum using the outputs fron the function tvar.

The inputs to the function are

arcoef: time-varying AR coefficients.

sigma2:   variance of the observation noise.

span: local stationary span.

nf: number of frequencies in computing specrum.

```
> data(MYE1F)
> # trend.order k=1
> z <- tvar(MYE1F, trend.order = 1, ar.order = 8, span = 20,
tau2.ini = 1.0e-03, delta = 1.0e-04,plot=F)
> spec <- tvspc(z$arcoef, z$sigma2, span=20, nf=400)
> plot(spec, tvv=v$tvv, dx=2, dy=0.10)
> #
> # trend.order k=2
> z <- tvar(MYE1F, trend.order = 2, ar.order = 8, span = 20,
tau2.ini = 6.6e-06, delta = 1.0e-06, plot=F)
> spec <- tvspc(z$arcoef, z$sigma2, span=20, nf=400)
> plot(spec, tvv=v$tvv, dx=2, dy=0.10)
```

Figure 13.3 illustrates the time-varying spectrum obtained from equation (13.20). The left plot shows the case for $k = 1$ and the right one the case for $k = 2$. In the plots of Figure 13.3, the horizontal and the vertical axes indicate the frequency and the logarithm of the spectra, respectively, and the slanted axis indicates time. From the figures, it can be seen that the power of the spectrum around $f = 0.25$ increases with the arrival of the P-wave. Subsequently, the power around $f = 0.1$ increases with the arrival of the S-wave. After that, the peaks of the spectra gradually shift to the right, and the spectrum eventually converges to that of the original background motions.

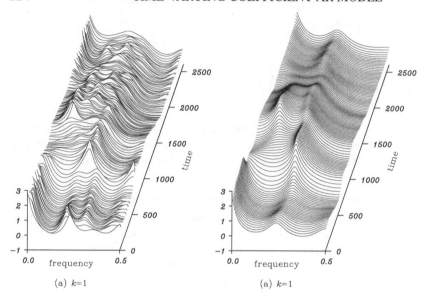

Figure 13.3: *Time-varying spectra of the seismic data.*

13.4 The Assumption on System Noise for the Time-Varying Coefficient AR Model

As stated in Section 13.2, the time-varying AR coefficients can be estimated by approximating the time-change of AR coefficients using the
trend component models, and then the TVCAR model can be expressed
in a state-space form. In the state-space model (13.17), the variance
covariance matrix for the system noise is assumed to be a diagonal form
given by $Q = \mathrm{diag}\left\{\tau^2, \cdots, \tau^2\right\}$. At first glance, this appears to be a very
strong unrealistic assumption. However, this section shows that it can
be obtained naturally by considering the smoothness of the frequency
response function of the AR operator.

Firstly, we consider the Fourier transform of the AR coefficients of
the TVCAR model,

$$A(f,n) = 1 - \sum_{j=1}^{m} a_{nj} e^{-2\pi i j f}, \qquad -\frac{1}{2} \leq f \leq \frac{1}{2}. \qquad (13.21)$$

This is the frequency response function of the AR model considered
as a whitening filter. Then the time-varying spectrum of (13.20) can be

expressed as

$$p_n(f) = \frac{\sigma^2}{|A(f,n)|^2}. \qquad (13.22)$$

Since the characteristics of the power spectrum are determined by the frequency response function $A(f,n)$, we can obtain good estimates of the time-varying spectrum by controlling the smoothness of $A(f,n)$. By considering the smoothness of $A(f,n)$ with respect to n, we can obtain the following models for the time change of the AR coefficients. Considering the k-th difference of the $A(f,n)$ to evaluate the smoothness of the time change of the AR coefficients, we obtain

$$\Delta^k A(f,n) = \sum_{j=1}^{m} \Delta^k a_{nj} e^{-2\pi i j f}. \qquad (13.23)$$

Then taking the integral of the square of $\Delta^k A(f,n)$, we obtain

$$\int_{-\frac{1}{2}}^{\frac{1}{2}} |\Delta^k A(f,n)|^2 df = \sum_{j=1}^{m} (\Delta^k a_{jn})^2. \qquad (13.24)$$

Therefore, it is possible to curtail the change over time of the power spectrum, by reducing the sum of the squares of the k-th differences of the AR coefficients. Since the squares of the k-th differences of the AR coefficients add up with the same weights in equation (13.24), this is equivalent to making the natural assumption that in equation (13.10),

$$\tau_{11}^2 = \cdots = \tau_{mm}^2 = \tau^2, \qquad (13.25)$$

and that

$$\Delta_n^k a_{jn} = v_{nj}, \qquad v_{nj} \sim N(0, \tau^2), \qquad j = 1, \cdots, m. \qquad (13.26)$$

13.5   Abrupt Changes of Coefficients

For the seismic data shown in Figure 13.1, it can be seen that the behavior of the wave form changes abruptly as a new signal arrives at a certain time. In this case, the estimated time-varying coefficients often become too variable or too smooth to capture the change of the characteristics as seen in Figure 13.2.

In such a case, by applying the locally stationary AR model shown in Chapter 8, we can obtain estimates of the change points. By using the estimates of arrival times of a new signal, it is possible to obtain better estimates of the time-varying AR coefficients. Assume that abrupt changes are detected at times $n = n_1, \ldots, n_p$. Corresponding to these points, the noise term v_{nj} of (13.10) takes a large negative or positive value. This indicates that it is necessary to increase the variance τ^2 at the time points $n = n_1, \ldots, n_p$.

Actualy, when τ^2 is increased in this way, the absolute value of the k-th time-difference of a_{nj} becomes larger. Thus, for $k = 1$, the coefficient a_{nj} shows a stepwise behavior and produces a discontinuous point. On the other hand, for $k = 2$, the slope of a_{nj} changes abruptly and yields a bending point. To yield a jump for a model with $k \geq 2$, it is necessary to add noise to each component of the state vector. Therefore we can realize the jump either by initializing the state and the variance covariance matrix $x_{n|n-1}$ and $V_{n|n-1}$ or by adding a large positive value to all diagonal elements of $V_{n|n-1}$ for $n = n_1, \ldots, n_p$. This method is applicable to trend estimation as well as for the time-varying coefficient modeling.

Example (Time-varying spectrum with abrupt structural changes)

To treat abrupt structural changes in the function `tvspec`, we specify the location of the change points by using the parameter `outlier` as follows:

```
> data(MYE1F)
> z <- tvar(MYE1F, trend.order = 2, ar.order = 8, span = 20,
outlier = c(630, 1026), tau2.ini = 6.6e-06, delta = 1.0e-06)
> spec <- tvspc(z$arcoef, z$sigma2)
> plot(spec, tvv=v$tvv, dx=2, dy=0.1)
```

Figure 13.4 shows the time-varying AR coefficients and the time-varying spectrum estimated by the model with $k = 2$ and $m = 4$ and by assuming that sudden structural changes occur at $n = 630$ and $n = 1026$ corresponding to the arrivals times of the P-wave and the S-wave, respectively. The estimated coefficients jump twice at these two points but change very smoothly elsewhere.

Moreover, from Figure 13.4, we can observe two characteristics of the series, firstly, that the dominant frequencies gradually shift to the right after the S-wave arrives, and secondly, that the specrum at direct current $p(0)$ increases and the shape of the spectrum goes back to that of the original background microtremors.

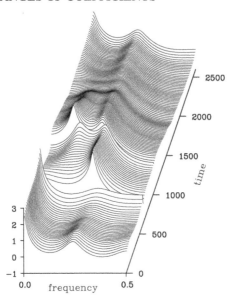

Figure 13.4 *Time-varying spectrum estimated by assuming discontinuities in the parameters.*

Problems

1. Give an example of a heteroscedastic (variance changing over time) time series other than earthquake data.
2. Consider a method of estimating a regression model whose coefficients change over time.

Chapter 14

Non-Gaussian State-Space Model

In this chapter, we extend the state-space model to cases where the system noise and/or the observation noise are non-Gaussian. This non-Gaussian model is applicable when there are sudden changes in the parameters caused by structural changes of the system or by outliers in the time series. For the general non-Gaussian models we consider here, the Kalman filtering and the smoothing algorithms sometimes cannot yield good estimates of the state. Even in such cases, however, we can derive an exact sequential formula for filtering and smoothing which can be implemented using numerical integration. Various applications of the non-Gaussian state-space model and an extension to nonlinear state-space model are also briefly shown.

14.1 Necessity of Non-Gaussian Models

As shown in the previous chapters, various types of time series models can be expressed in terms of the linear-Gaussian state-space model

$$
\begin{aligned}
x_n &= F x_{n-1} + G v_n \\
y_n &= H x_n + w_n,
\end{aligned}
\tag{14.1}
$$

where y_n is the time series, x_n is the unknown state vector, and v_n and w_n are Gaussian white noises.

The state-space model is a very useful tool for time series modeling, since it is a natural representation of a time series model and provides us with very efficient computational methods such as Kalman filtering and smoothing algorithms. Although many important time series models are expressible as linear-Gaussian state-space models, there are some other situations where extensions of the model are necessary, such as the case of a nonstationary time series with time-varying stochastic structure that sometimes contains both smooth and abrupt changes. While a linear-Gaussian state-space model can reasonably express gradual structural changes of nonstationary time series, it is necessary to build a complex model to properly address abrupt changes. To remove the influence of outliers in the data, development of an automatic detection method for the outliers or a robust estimation procedure is necessary (West (1981), Meinhold and Singpurwalla (1989) and Carlin et al. (1992)). In addition, nonlinear systems and discrete processes cannot be adequately modeled by standard linear Gaussian state-space models.

Let us consider possible solutions to these problems. In state-space modeling, changes in the stochastic structure are usually reflected by changes in the state. Assuming a heavy-tailed distribution such as the Cauchy distribution for the system noise v_n, both smooth changes that occur with high probability and abrupt changes that occur with low probability can be expressed by a single noise distribution.

Similarly, it is reasonable to deal with outliers in time series by using a heavy-tailed distribution for the observation noise w_n. On the other hand, if the system contains nonlinearity or if the observations are discrete values, the distribution of the state vector inevitably becomes non-Gaussian.

Therefore, as a key to the solution of problems that occur with standard state-space modeling, a treatment of non-Gaussian state distributions is essential. In the following sections, the recursive filter and smoothing algorithms for the estimation of the unknown states of non-Gaussian state-space models and their applications are presented.

14.2   Non-Gaussian State-Space Models and State Estimation

Consider the following state-space model

$$
\begin{aligned}
x_n &= Fx_{n-1} + Gv_n & \text{(system model)} & \quad (14.2) \\
y_n &= Hx_n + w_n & \text{(observation model),} & \quad (14.3)
\end{aligned}
$$

where the system noise v_n and the observation noise w_n are white noises that follow the density functions $q(v)$ and $r(w)$, respectively. In contrast to the state-space models presented in the previous chapters, these distributions are not necessarily Gaussian.

In this case, the distribution of the state vector x_n generally becomes non-Gaussian. Consequently, the state-space model of equations (14.2) and (14.3) is called a *non-Gaussian state-space model*. Clearly, this non-Gaussian state-space model is an extension of the standard state-space model.

As in Chapter 9, the information from the time series obtained until the time j is denoted as $Y_j \equiv \{y_1, \ldots, y_j\}$. Similarly, the set of realizations of the state x_n up to time j is denoted as $X_j \equiv \{x_1, \ldots, x_j\}$. Further, the initial state vector x_0 is assumed to be distributed according to the probability density function $p(x_0|Y_0)$. Here the state estimation problem is to obtain the conditional distribution of the state vector x_n, given the information Y_m. There are three versions of the state estimation problem, depending on the relation between n and m. Specifically, for $n > m$, $n = m$ and $n < m$, it is called the prediction problem, the filtering problem and the smoothing problem, respectively.

For the linear-Gaussian state-space model, the Kalman filter provides a recursive algorithm for obtaining the conditional mean and the conditional variance covariance matrix of the state vector. On the other hand, with the general non-Gaussian state-space model, the posterior state distribution cannot be completely determined by these quantities, and it is necessary to obtain the conditional densities for state estimation.

However, for the state-space models defined by (14.2) and (14.3), by using the relations $p(x_n|x_{n-1}, Y_{n-1}) = p(x_n|x_{n-1})$ and $p(y_n|x_n, Y_{n-1}) = p(y_n|x_n)$, we can derive recursive formulas for obtaining the one-step-ahead predictive distribution $p(x_n|Y_{n-1})$ and the filtering distribution $p(x_n|Y_n)$ as follows (Kitagawa (1987)). Note that the following algorithm can be applied to general nonlinear state-space models (Kitagawa (1991), Kitagawa and Gersch (1996)).

[One-step-ahead prediction]

$$
\begin{aligned}
p(x_n|Y_{n-1}) &= \int_{-\infty}^{\infty} p(x_n, x_{n-1}|Y_{n-1}) dx_{n-1} \\
&= \int_{-\infty}^{\infty} p(x_n|x_{n-1}, Y_{n-1}) p(x_{n-1}|Y_{n-1}) dx_{n-1} \\
&= \int_{-\infty}^{\infty} p(x_n|x_{n-1}) p(x_{n-1}|Y_{n-1}) dx_{n-1}, \quad (14.4)
\end{aligned}
$$

[Filtering]

$$
\begin{aligned}
p(x_n|Y_n) &= p(x_n|y_n, Y_{n-1}) \\
&= \frac{p(y_n|x_n, Y_{n-1}) p(x_n|Y_{n-1})}{p(y_n|Y_{n-1})} \\
&= \frac{p(y_n|x_n) p(x_n|Y_{n-1})}{p(y_n|Y_{n-1})}, \quad (14.5)
\end{aligned}
$$

where $p(y_n|Y_{n-1})$ is obtained as $\int p(y_n|x_n) p(x_n|Y_{n-1}) dx_n$. The one-step-ahead prediction formula (14.4) is an extension of the one-step-ahead prediction of the Kalman filter. Here $p(x_n|x_{n-1})$ is the density function of the state x_n when the previous state x_{n-1} is given, which is determined by the system model (14.2). Therefore, if the filter $p(x_{n-1}|Y_{n-1})$ of x_{n-1} is given, the one-step-ahead predictor $p(x_n|Y_{n-1})$ can be evaluated. On the other hand, the filter formula (14.5) is an extension of the filtering step of the Kalman filter. $p(y_n|x_n)$ is the conditional distribution of the observation y_n, when the state x_n is given. It is determined by the observation model of (14.3). Therefore, if the predictive distribution $p(x_n|Y_{n-1})$ of x_n is given, then the filter density $p(x_n|Y_n)$ is computable.

Next, we consider the smoothing problem. Using the equation $p(x_n|x_{n+1},Y_N) = p(x_n|x_{n+1},Y_n)$ that holds for the state-space models of (14.2) and (14.3), we obtain

$$
\begin{aligned}
p(x_n,x_{n+1}|Y_N) &= p(x_{n+1}|Y_N)p(x_n|x_{n+1},Y_N) \\
&= p(x_{n+1}|Y_N)p(x_n|x_{n+1},Y_n) \\
&= p(x_{n+1}|Y_N)\frac{p(x_n|Y_n)p(x_{n+1}|x_n,Y_n)}{p(x_{n+1}|Y_n)} \\
&= p(x_{n+1}|Y_N)\frac{p(x_n|Y_n)p(x_{n+1}|x_n)}{p(x_{n+1}|Y_n)}.
\end{aligned}
\tag{14.6}
$$

Integration of both sides of (14.6) yields the following sequential formula for the smoothing problem:

[Smoothing formula]

$$
\begin{aligned}
p(x_n|Y_N) &= \int_{-\infty}^{\infty} p(x_n,x_{n+1}|Y_N)dx_{n+1} \\
&= p(x_n|Y_n)\int_{-\infty}^{\infty} \frac{p(x_{n+1}|Y_N)p(x_{n+1}|x_n)}{p(x_{n+1}|Y_n)}dx_{n+1}.
\end{aligned}
\tag{14.7}
$$

In the right-hand side of the formula (14.7), $p(x_{n+1}|x_n)$ is determined by the system model (14.2). On the other hand, $p(x_n|Y_n)$ and $p(x_{n+1}|Y_n)$ are obtained by equations (14.4) and (14.5), respectively. Thus the smoothing formula (14.7) indicates that if $p(x_{n+1}|Y_N)$ is given, we can compute $p(x_n|Y_N)$.

Since $p(x_N|Y_N)$ can be obtained by filtering (14.5), by repeating the smoothing formula (14.7) for $n = N-1,\dots,1$ in a similar way as for the fixed interval smoothing presented in Chapter 9, we can obtain the smoothing distributions $p(x_{N-1}|Y_N),\dots,p(x_1|Y_N)$, successively.

14.3   Numerical Computation of the State Estimation Formula

As shown in the previous section, we can derive recursive estimation formulas for the non-Gaussian state-space model that are natural extensions of the Kalman filter. Adopting this comprehensive algorithm, we can extensively treat various types of time series models. However, in the practical application of the algorithm, difficulties often arise in calculating the formulas of the filter and smoother.

For a linear Gaussian state-space model, all the conditional distributions $p(x_n|Y_{n-1})$, $p(x_n|Y_n)$ and $p(x_n|Y_N)$ become normal distributions. Therefore, in that case, only the mean vectors and the variance covariance matrices need to be evaluated, and correspondingly (14.4), (14.5) and (14.7) become equivalent to the ordinary Kalman filter and the smoothing algorithms. However, since the conditional distribution $p(x_n|Y_j)$ of the state generally becomes a non-Gaussian distribution, it cannot be specified using only the mean vector and the variance covariance matrix. Various algorithms have been presented, for instance, the extended Kalman filter (Anderson and Moore (1979)) and the second order filter, to approximate the non-Gaussian distribution by a single Gaussian distribution with properly determined mean vector and variance covariance matrix. In general, however, they do not perform well.

This section deals with the method of realizing the non-Gaussian filter and the non-Gaussian smoothing algorithm by numerically approximating the non-Gaussian distributions (Kitagawa (1987)). In this approach, a non-Gaussian state density function is approximated numerically using functions such as a step function, a piecewise linear function or a spline function. Then the formulas (14.4) – (14.7) can be evaluated by numerical computation. Since this approach requires a huge amount of computation, it used to be considered an impractical method. Nowadays, with the development of high-speed computers, those numerical methods have become practical, at least for low-dimensional systems. In this section, we approximate the density functions that appeared in (14.4), (14.5) and (14.7) by simple step functions (Kitagawa and Gersch (1996)).

To be specific, the density function $f(t)$ to be approximated is defined on a line: $-\infty < t < \infty$. To approximate this density function by a step function, the domain of the density function is firstly restricted to a finite interval $[t_0, t_d]$, which is then divided into d subintervals $t_0 < t_1 < \cdots < t_d$. Here t_0 and t_d are assumed to be sufficiently small and large numbers, respectively. In the actual programming, we change t_0 and t_i depending on the location of the density function. However, for simplicity, ends of the sub-intervals are assumed to be fixed in the following. Further, the width of the subintervals is assumed to be identical. Then the i-th point is given by $t_i = t_0 + i\Delta t$ with $\Delta t = (t_d - t_0)/d$.

Table 14.1: *Approximation of density functions.*

density function	approximation	denotation
$p(x_n\|Y_{n-1})$	$\{d;t_0,\cdots,t_d;p_1,\cdots,p_d\}$	$\tilde{p}(t)$
$p(x_n\|Y_n)$	$\{d;t_0,\cdots,t_d;f_1,\cdots,f_d\}$	$\tilde{f}(t)$
$p(x_n\|Y_N)$	$\{d;t_0,\cdots,t_d;s_1,\cdots,s_d\}$	$\tilde{s}(t)$
$q(v)$	$\{2d+1;t_{-d},\cdots,t_d;q_{-d},\cdots,q_d\}$	$\tilde{q}(v)$

In a step-function approximation, the function $f(t)$ is approximated by f_i on the subinterval $[t_{i-1},t_i]$. If the function $f(t)$ is actually a step-function, it is given by $f_i = f(t_i)$. But in general, it may be defined by

$$f_i = \Delta t \int_{t_{i-1}}^{t_i} f(t)dt. \qquad (14.8)$$

Using those values, the step-function approximation of the function $f(t)$ is specified by $\{d;t_0,\ldots,t_d;f_1,\ldots,f_d\}$. In the following, the approximated function is denoted by $\tilde{f}(t)$. For the numerical implementation of the non-Gaussian filter and the smoothing formula, it is necessary to approximate the density functions $p(x_n|Y_{n-1})$, $p(x_n|Y_n)$, $p(x_n|Y_N)$ and the system noise density $q(v)$ as shown in Table 14.1. On the other hand, the observation noise density $r(v)$ can be used directly without discretizing it. Alternative approaches for numerical integrations are Gauss-Hermite polynomial integration (Schnatter (1992)), a random replacement of knots spline function approximation (Tanizaki (1993)) and Monte Carlo integration (Carlin et al. (1992), Frühwirth-Schnatter (1994), Carter and Kohn (1993)).

In the following, we show a procedure for numerical evaluation of the simplest one-dimensional trend model;

$$
\begin{aligned}
x_n &= x_{n-1} + v_n, \\
y_n &= x_n + w_n.
\end{aligned} \qquad (14.9)
$$

[One-step-ahead prediction]

For $i = 1,\ldots,d$, compute

$$
\begin{aligned}
p_i = \tilde{p}(t_i) &= \int_{t_0}^{t_d} \tilde{q}(t_i - s)\tilde{f}(s)ds \\
&= \sum_{j=1}^{d} \int_{t_{j-1}}^{t_j} \tilde{q}(t_i - s)\tilde{f}(s)ds
\end{aligned}
$$

$$= \Delta t \sum_{j=1}^{d} q_{i-j} f_j. \tag{14.10}$$

[Filtering]

For $i = 1, \ldots, d$, compute

$$f_i = \tilde{f}(t_i) = \frac{r(y_n - t_i)\tilde{p}(t_i)}{C} = \frac{r(y_n - t_i)p_i}{C}, \tag{14.11}$$

where the normalizing constant C is obtained by

$$C = \int_{t_0}^{t_d} r(y_n - t)\tilde{p}(t)dt = \sum_{j=1}^{d} \int_{t_{j-1}}^{t_j} r(y_n - t)\tilde{p}(t)dt$$

$$= \Delta t \sum_{j=1}^{d} r(y_n - t_j)p_j. \tag{14.12}$$

[Smoothing]

For $i = 1, \ldots, d$, compute

$$s_i = \tilde{s}(t_i) = \tilde{f}(t_i) \int_{t_0}^{t_d} \frac{\tilde{q}(t_i - u)\tilde{s}(u)}{\tilde{p}(u)}du$$

$$= \tilde{f}(t_i) \sum_{j=1}^{d} \int_{t_{j-1}}^{t_j} \frac{\tilde{q}(t_i - u)\tilde{s}(u)}{\tilde{p}(u)}du$$

$$= \Delta t \cdot f(t_i) \sum_{j=1}^{d} \frac{q_{i-j}s_j}{p_j}. \tag{14.13}$$

It should be noted that, in practical computation, even after the prediction step (14.10) and the smoothing (14.13), the density functions should be normalized so that the value of the integral over the whole interval becomes 1. This can be achieved, for example, by modifying f_i to $f_i/I(f)$ where $I(f)$ is defined by

$$I(f) = \int_{t_0}^{t_d} f(t)dt = \Delta t\,(f_1 + \cdots + f_d). \tag{14.14}$$

14.4 Non-Gaussian Trend Model

In the trend model considered in Chapter 11, in cases where the distribution of the system noise v_n or the observation noise w_n is non-Gaussian,

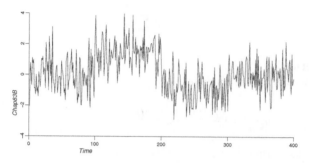

Figure 14.1: *Data generated by the model of (14.15).*

we obtain a *non-Gaussian trend model*. In this section, numerical examples are given in order to explain the features of the non-Gaussian model.

Example (Non-Gaussian trend estimation)

The R function `ngsmth` of the package `TSSS` performs non-Gaussian filtering and smoothing for one-dimensional state-space model with non-Gaussian noise densities. The inputs to the function are

y:	a univariate time series.
noisev:	type of system noise density.
	1: Gaussian (normal) distribution,
	2: Type VII Pearson family of distribution.
	3: two-sided exponential distribution.
tau2	dispersion of system noise.
bv:	shape parameter of the sytem noise.
noisew:	type of system noise density.
	1: Gaussian (normal) distribution,
	2: Type VII Pearson family of distribution.
	3: two-sided exponential distribution.
	4: double exponential distribution.
sigma2:	dispersion of the observation noise
bw:	shape parameter of the obserbaip noise.
initd:	type of initial distribution.
	1: Gaussian (normal) distribution,
	2: uniform distribution.
	3: two-sided exponential distribution.
k:	number of intervals
plot:	logical. If TRUE (default), trend and smt are plotted

and the outputs from this function are:

trend: 7 percentile points of smoothed posterior distribution of the trend.

smt: smoothed density for $n = 1, ..., N$.

```
> data(PfilterSample)
> y <- PfilterSample
> par(mar=c(3,3,1,1)+0.1)
> #
> # system noise density : Gaussian (normal)
> s1 <- ngsmth(y, noisev = 1, tau2 = 1.4e-02, noisew = 1,
sigma2 = 1.048)
> # 3D plot
> plot(s1, "smt", theta = 25, phi = 30, expand = 0.25,
col="white")
> #
> # system noise density : Pearson family with b=0.6
> s2 <- ngsmth(y, noisev = 2, tau2 = 2.11e-10, bv = 0.6,
noisew = 1, sigma2 = 1.042)
> plot(s2, "smt", theta = 25, phi = 30, expand =
0.25,col="white")
> #
> # system noise density : Cauchy (Pearson family with b=1)
> s2 <- ngsmth(y, noisev = 2, tau2 = 3.53e-5, bv = 1.0,
noisew = 1,sigma2 = 1.045)
```

Figure 14.1 shows the test data generated by the following models;

$$y_n \sim N(\mu_n, 1), \qquad \mu_n = \begin{cases} 0, & 1 \leq n \leq 100 \\ 1, & 101 \leq n \leq 200 \\ -1, & 201 \leq n \leq 300 \\ 0, & 301 \leq n \leq 400 \end{cases}. \qquad (14.15)$$

In the plot, it can be confirmed that the mean value of the data abruptly changes three times. For estimation of a changing mean value function with the structural changes shown in Figure 14.1, we consider the following first order trend model;

$$\begin{aligned} t_n &= t_{n-1} + v_n \\ y_n &= t_n + w_n, \end{aligned} \qquad (14.16)$$

where the observation noise w_n is assumed to be Gaussian, and the system noise v_n follows the type VII Pearson family of distributions:

$$q(v_n) = \frac{c}{(\tau^2 + v_n^2)^b}, \qquad \frac{1}{2} < b < \infty. \qquad (14.17)$$

Table 14.2 *Non-Gaussian model with Pearson family of distributions with various values of shape parameter b.*

b	τ^2	σ^2	log-likelihood	AIC
0.60	0.211×10^{-9}	1.042	-597.19	1198.38
0.75	0.299×10^{-7}	1.043	-597.39	1198.78
1.00	0.353×10^{-4}	1.045	-597.99	1198.98
1.50	0.303×10^{-2}	1.045	-599.13	1202.26
3.00	0.406×10^{-1}	1.046	-600.40	1204.80
∞	0.140×10^{-1}	1.048	-600.69	1205.38

Here b and τ^2 are the shape parameter and the dispersion parameter, respectively. c is a normalizing constant which makes the integral of $q(v)$ over the whole interval equal to 1 and is given by $c = \tau^{2b-1}\Gamma(b)/\Gamma(\frac{1}{2})\Gamma(b-\frac{1}{2})$ (Johnson and Kotz (1970)). The Pearson family of distributions can express various symmetric probability density functions with heavier-tailed distributions than the normal distribution. For example, the Pearson family of distributions yields the Cauchy distribution for $b = 1$, the t distribution with k degrees of freedom for $b = (k+1)/2$ and the normal distribution as $b \to \infty$.

Table 14.2 summarizes the values of the maximum log-likelihoods and the AIC's of the Pearson family of distributions with $b = 3/k$, ($k = 0,\dots,5$). The table shows that the AIC is minimized at $b = 0.60$, and the AIC of the normal distribution model ($b = \infty$) is the maximum.

The plot on the left-hand side of Figure 14.2 shows the change over time of the smoothed distribution of the trend $p(t_n|Y_N)$, obtained from the Gaussian model for ($b = \infty$), and the one on the right-hand side shows the corresponding plot obtained from the non-Gaussian model with $b = 0.6$. The left plot shows that the distribution of the trend estimated by the Gaussian model gradually shifts left or right with the progress of time n. On the other hand, for the case of the non-Gaussian model shown in the right plot, the estimated density is seen to be very stable with abrupt changes at three time points.

In Figure 14.3, the plot on the left-hand side shows the mean and the $\pm 1,2,3$ standard deviation intervals of the estimated distribution at each time point for the Gaussian model. On the other hand, the plot on the right-hand side shows the $0.13, 2.27, 15.87, 50.0, 84.13, 97.73$ and 99.87 percentile points of the estimated trend distribution that correspond to the mean and $\pm1, 2, 3$ intervals of the Gaussian distribution. Comparing the two plots in Figure 14.3, it can be seen that the non-Gaussian model yields a smoother estimate than the Gaussian model and clearly detects jumps in the trend.

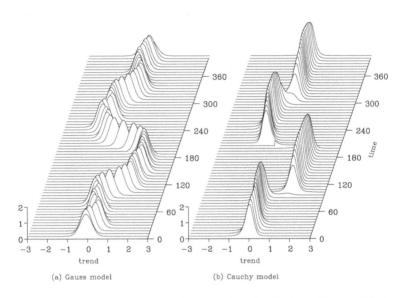

(a) Gauss model   (b) Cauchy model

Figure 14.2 *Changes over time of the smoothed distribution of the trend. Left: Gaussian model, right: non-Gaussian model.*

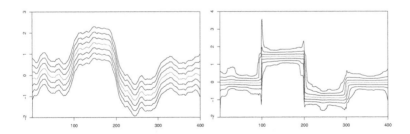

Figure 14.3 *Estimation of trend by Gaussian (left plot) and Pearson distribution with $b = 0.6$ (right plot) models.*

14.5   Non-symmetric Distribution – A Time-Varying Variance Model

The distribution of data observed in the real world sometimes shows non-symmetry. This is due to characteristics such as nonlinearity or non-Gaussianity of the data-generating mechanism. If we apply the least-squares method to such a data set, we may obtain a biased estimate of the trend. Thus, if the distribution of the observations is known, we can obtain better estimates by using the non-Gaussian distribution $r(w)$ for filtering and smoothing. In this section, for the estimation of the time-varying variance, we re-consider the model treated in Section 13.1;

$$t_m = t_{m-1} + v_m$$
$$y_m = t_m + w_m. \tag{14.18}$$

The R function ngsmth of the package TSSS is also applicable for the filtering and smoothing with non-Gaussian state-space model with non-symmetric noise distribution.

```
> data(MYE1F)
> n <- length(MYE1F)
> # data transofrmation
> yy <- rep(0, n)
> for (i in 2:n) yy[i] <- MYE1F[i] - 0.5 * MYE1F[i-1]
> m <- seq(1, n, by = 2)
> y <- yy[m]
> z <- tvvar(y, trend.order = 2, tau2.ini = 4.909e-02, delta =
1.0e-06)
>
> par(mar=c(2,2,1,1)+0.1)
> plot(z$sm,type="l")
>
> # system noise: Gaussian, observation noise: Guassian
> s3 <- ngsmth(z$sm, noisev = 1, tau2 = z$tau2, noisew = 1,
sigma2 = pi*pi/6, k = 190)
```

Example   In Chapter 13, the double exponential distribution for the observation noise w_n is approximated by a Gaussian distribution $N(-\zeta, \pi^2/6)$. Here we shall compare the smoothing results obtained by this Gaussian approximation with the one based on the exact double-exponential distribution model using non-Gaussian filter and smoother.

Figure 14.4 shows the transformed seismic data obtained from the original data of Figure 1.1(f). The plotted data set was obtained by sam-

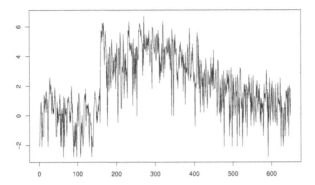

Figure 14.4: *Transformed seismic data.*

pling every two points after filtering by $\tilde{y}_n = y_n - 0.5y_{n-1}$ and transformed by the method discussed in Section 13.1. In this chapter, since it is assumed that the time series is a white noise, this transformation was applied in order to weaken the strong autocorrelation that can be observed in the first part of the data.

Fitting the Gaussian model with $\sigma^2 = \pi^2/6$ to the data, we obtain the maximum likelihood estimate $\hat{\tau}^2 = 0.04909$ and AIC $= 2165.10$. Figure 14.5 shows the wiggly trend estimated by this model. This is due to the large deviations to the negative side, which occured due to the log transformation shown in Figure 14.4, which strongly affect the estimated trend.

On the other hand, the maximum likelihood estimates of the non-Gaussian model with a Cauchy distribution for the system noise and a double exponential distribution for the observation noise

$$
\begin{aligned}
q(x) &= \frac{1}{\pi} \frac{\tau}{\tau^2 + x^2} \\
r(x) &= \exp\{x - e^x\},
\end{aligned}
\tag{14.19}
$$

are $\hat{\tau}^2 = 0.000260$ and AIC $= 2056.97$, respectively.

```
> # observation noise: double exponential, system noise: Cauchy
> s3 <- ngsmth(z$sm, noisev = 2, tau2 = 2.6e-4, noisew = 4,
sigma2 = 1.0, k = 190)
```

From the big differences between the AIC values of the two models, it is evident that the non-Gaussian model is much better than the Gaus-

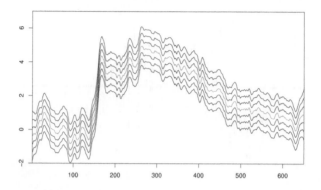

Figure 14.5: *Estimated trend (log-variance) by Gaussian model.*

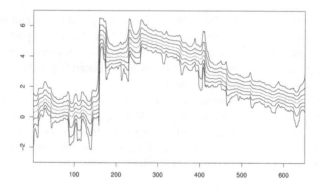

Figure 14.6: *Estimated trend (log-variance) by a non-Gaussian model.*

sian model. Actually, as shown in Figure 14.6, the trend obtained with the non-Gaussian model is smoother by far than the one obtained by the Gaussian model and clearly detected sudden increases in the variance due to the arrivals of the P-wave and the S-wave.

Moreover, the plot of this model shows that the observations with a large downward deviation influenced only the curve of 0.13 percent points and not that of the 50 percent points. In summary, by non-Gaussian smoothing with an appropriate model, we can obtain an estimate of the trend that is able to automatically detect a jump of the parameter but is not unduly influenced by a non-symmetric distribution. This estimation method of a time-varying variance can be also used for the estimation of the stochastic volatility of a financial time series as shown below.

Example (Stock price index data)

```
> data( Nikkei225 )
> y <- diff( log(Nikkei225) )
> z <- tvvar( y, trend.order = 2, tau2.ini = 1.6e-04, delta=
1.0e-06)
> par( mar=c(2,2,1,1)+0.1, mfrow=c(2,1) )
> plot( z$sm, type="l",xaxp=c(0,700,7) )
> s <- ngsmth(z$sm, noisev = 1, tau2 = 6.9e-2, noisew = 4,
sigma2 = 1.0, k = 190) > s2 <- s$trend
> s3 <-exp(s2[,2]/2 )
> plot( s3,type="l",ylim=c(0,0.03),xaxp=c(0,700,7) )
```

Figure 14.7 (a) shows the transformed series s_m obtained by applying the method shown in Section 13.1 to the difference of the logarithm of Nikkei 225 stock price index data. On the other hand, Figure 14.7(b) shows the posterior distribution of the estimate of $\log \sigma_n^2$ obtained as the trend of this series estimated by the non-Gaussian model of (14.18). In addition, an estimate of the volatility is obtained from the exponential of a half of the central (the 50 percentile) curve in plot (b), i.e., by

$$\exp\left(\log \hat{\sigma}_n^2/2\right) = \sqrt{\hat{\sigma}_n^2} = \hat{\sigma}_n.$$

This method of estimating time-varying variance can be immediately applied to the smoothing of the periodogram. Since the periodogram follows the χ^2 distribution with 2 degrees of freedom, the series obtained by taking the logarithm of the periodogram thus follows the same distribution as given in (14.19). Therefore, the method treated in this chapter can be applied to smoothing of the logarithm of the periodogram as well.

Moreover, using a non-Gaussian model, it is possible to estimate the time-varying variance from the transformed data

$$z_n = \log y_n^2. \tag{14.20}$$

In this case, for the density function of the observation noise, we use the density function obtained as the logarithm of the χ^2 distribution with 1 degree of freedom, i.e.,

$$r(w) = \frac{1}{\sqrt{2\pi}} \exp\left\{\frac{w - e^w}{2}\right\}. \tag{14.21}$$

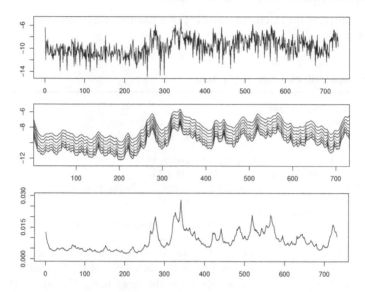

Figure 14.7 *(a) Transformed Nikkei 225 stock price index data, (b) posterior distribution of* $\log \sigma_n^2$, *(c) estimates of volatility.*

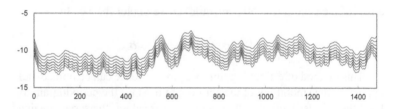

Figure 14.8 *The smoothed distribution obtained by the model of (14.21) for the Nikkei 225 stock price index data.*

Here the time-varying variance can be estimated without making the assumption that $\log \sigma_{2m-1}^2 = \log \sigma_{2m}^2$. Figure 14.8 shows the estimate of $\log \sigma_n^2$ obtained by applying this method to the logarithm of the differences of the Nikkei 225 stock price index data.

14.6 Applications of the Non-Gaussian State-Space Model

In this section, several applications of the non-Gaussian state-space model are presented. Specifically, modeling for processing outliers, smoothing for discrete processes and a direct method for estimating time-varying variances are shown. Extension to nonlinear state-space modeling is also shown.

14.6.1 Processing of the outliers by a mixture of Gaussian distributions

Here we consider treatment of outliers that follow a distribution different from the usual noise distribution of observations. To treat outliers in a time series, assuming that the outliers appear with low probability $\alpha \ll 1$, it will be reasonable to use a Gaussian mixture distribution for the observation noise w_n,

$$r(w) = (1 - \alpha)\varphi(w; \mu, \sigma^2) + \alpha\varphi(w; \xi, \delta^2). \qquad (14.22)$$

Here $\varphi(x; \mu, \sigma^2)$ denotes the density function of the normal distribution with mean μ and variance σ^2. The first and second terms on the right-hand side express the distributions of the normal observations and the outliers, respectively. In particular, by putting $\xi = \mu = 0$ and $\delta^2 \gg \sigma^2$, the observation noise of the outliers is assumed to have zero mean and a large variance. The Gaussian-sum filter and smoother for this Gaussian-mixture noise model is given in Alspach and Sorenson (1972) and Kitagawa (1989,1994).

The Cauchy distribution may be considered as an approximation of the above-mentioned mixture distribution. Therefore, the outliers can be treated, for example, using the Cauchy distribution or the Pearson family of distributions, which can be used to generate a density function of the observation noise. On the other hand, if outliers always appear on the positive side of the time series, as shown in Figure 1.1(h) of the ground-water data, we can estimate the trend properly, ignoring the outliers by setting $\xi \gg \mu$ and $\delta^2 = \sigma^2$.

14.6.2 A nonstationary discrete process

Non-Gaussian models can also be applied to the smoothing of discrete processes, for instance, nonstationary binomial distributions and Poisson distributions. Consider the time series $m_1, \ldots, m_N$ that takes integer values. Assume that the probability of taking value m_n at time n is given by the Poisson distribution

$$P(m_n | p_n) = \frac{e^{-p_n} p_n^{m_n}}{m_n!}, \qquad (14.23)$$

where p_n denotes the expected frequency of the occurrences of a certain event at a unit time. In this model, it is implicitly assumed that the mean of the Poisson distribution is not a constant but gradually changes with time. Such a model is called a *nonstationary Poisson process model*. As a model for the change over time of the Poisson mean, p_n, we may use the trend component model $p_n = p_{n-1} + v_n$. Here, since p_n is required to be positive, $0 < p_n < \infty$, we define a new variable $\theta_n = \log p_n$ and apply a trend component model for θ_n. Therefore, the model for smoothing the nonstationary Poisson process becomes

$$\theta_n = \theta_{n-1} + v_n, \tag{14.24}$$

$$P(m_n|\theta_n) = \frac{e^{-p_n} p_n^{m_n}}{m_n!}, \tag{14.25}$$

where p_n is defined by $p_n = e^{\theta_n}$. The first and second models correspond to the system model and the observation model, and yield conditional distributions of θ_n and m_n, respectively. Therefore the non-Gaussian filtering method presented in this chapter can be applied to the discrete process, and we can estimate the time-varying Poisson mean value from the data.

Similarly, the non-Gaussian state-space model can be applied to the smoothing of an inhomogeneous binary process. Assume that the probability of a certain event occurring m_n times in the ℓ_n observations at time n is given by the binomial distribution

$$P(m_n|p_n, \ell_n) = \binom{\ell_n}{m_n} p_n^{m_n} (1-p_n)^{\ell_n - m_n}. \tag{14.26}$$

Here, p_n is called the binomial probability, and expresses the expected number of occurrences of a certain event. In this case, if p_n varies with time, this model is called the *inhomogeneous binary process model*. p_n can be estimated by the following model:

$$\theta_n = \theta_{n-1} + v_n,$$

$$P(m_n|\theta_n, \ell_n) = \binom{\ell_n}{m_n} p_n^{m_n} (1-p_n)^{\ell_n - m_n}, \tag{14.27}$$

where $p_n = e^{\theta_n}/(1+e^{\theta_n})$.

14.6.3   A direct method of estimating the time-varying variance

In Section 14.5, the time-varying variance was estimated by smoothing the logarithm of the square of the original time series. However, it is also possible to estimate the time-varying variance directly using the following model:

$$
\begin{aligned}
t_n &= t_{n-1} + v_n, \\
y_n &\sim N(0, e^{t_n}).
\end{aligned}
\tag{14.28}
$$

Since the variance always takes a positive value in this model, the state t_n can be taken as the logarithm of the variance. Here, $\exp(t_n)$ becomes an estimate of the variance at time n. From the estimation of the time-varying variance of the Nikkei 225 stock price index data, we can obtain equivalent results to Figure 4.8. Here the obtained log-likelihood values for each model are different. This is because the data sets used in computing the log-likelihood are different. As shown in Section 4.6, we can confirm that these two models have exactly the same AIC values by compensating for the effect of data transformation by evaluating the Jacobian of the transformation.

14.6.4   Nonlinear state-space models

Consider a system described by a nonlinear state-space model

$$
\begin{aligned}
x_n &= f(x_{n-1}) + w_n \\
y_n &= h(x_n) + \varepsilon_n,
\end{aligned}
\tag{14.29}
$$

where y_n and x_n are one-dimensional time series and m-dimensional state vectors, respectively, and $f(x)$ and $h(x)$ are nonlinear functions of the state x. w_n and ε_n are m-dimensional and one-dimensional white noise sequences having densities $q(w)$ and $r(\varepsilon)$, respectively, which are independent of the past history of x_n and y_n. The initial state vector x_0 is assumed to be distributed according to the density $p(x_0)$.

For the nonlinear state-space model (14.29), the conditional distributions $p(x_n|x_{n-1})$ and $p(y_n|x_n)$ are given by

$$
\begin{aligned}
p(x_n|x_{n-1}) &= q(x_n - f(x_{n-1})) \\
p(y_n|x_n) &= r(y_n - h(x_n)).
\end{aligned}
\tag{14.30}
$$

Further, it can be verified that the following relations hold:

$$
\begin{aligned}
p(x_n|x_{n-1}, Y_{n-1}) &= p(x_n|x_{n-1}) \\
p(y_n|x_n, Y_{n-1}) &= p(y_n|x_n) \\
p(x_n|x_{n+1}, Y_N) &= p(x_n|x_{n+1}, Y_n).
\end{aligned}
\tag{14.31}
$$

Since the one-step-ahead prediction, filtering and smoothing formulae are obtained by using only the relations in (14.32), the non-Gaussian filtering and smoothing methods can be also applied for nonlinear state-space model. The detail of this method can be seen in Kitagawa (1991) and Kitagawa and Gersch (1996).

Problems

1. Discuss the advantages of using the Cauchy distribution or the two-sided exponential (Laplace) distribution in time series modeling.

2. Verify that when y_n follows a normal distribution with mean 0 and variance 1, the distribution of $v_n = \log y_n^2$ is given by (14.21).

3. Consider a model for which the trend follows a Gaussian distribution with probability α and is constant with probability $1 - \alpha$.

Chapter 15

Particle Filter

In this chapter, we consider a sequential Monte Carlo method of filtering and smoothing for a nonlinear non-Gaussian state-space model, which is usually called a particle filter and smoother. The particle filter and smoother have been developed as a practical and easily implementable method of filtering and smoothing for high-dimensional nonlinear non-Gaussian state-space models. In the particle filter, an arbitrary non-Gaussian distribution is approximated by many particles that can be considered to be independent realizations of the distribution. Then recursive filtering and smoothing algorithms are realized by three simple manipulations of particles, namely, substitution of particles into models, computation of importance of each particle and the re-sampling of the particles (Gordon et al. (1993), Kitagawa (1996), Doucet et al. (2001)).

15.1 The Nonlinear Non-Gaussian State-Space Model and Approximations of Distributions

In this section, we consider the following nonlinear non-Gaussian state-space model

$$x_n = F(x_{n-1}, v_n) \qquad (15.1)$$
$$y_n = H(x_n, w_n), \qquad (15.2)$$

where y_n is a time series, x_n is the k-dimensional state vector, and the models of (15.1) and (15.2) are called the system model and the observation model, respectively (Kitagawa and Gersch (1996)).

This model is a generalization of the state-space model that has been treated in the previous chapters. The system noise v_n and the observation noise w_n are assumed to be ℓ-dimensional and one-dimensional white noises that follow the density functions $q(v)$ and $r(w)$, respectively. The initial state vector x_0 is assumed to follow the density function $p_0(x)$. In general, the functions F and H are assumed to be nonlinear functions. For the function $y = H(x, w)$, assuming that x and y are given, w is

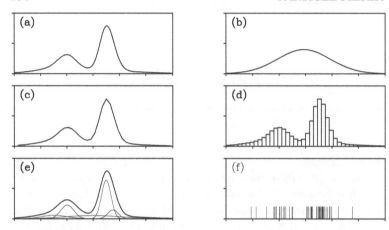

Figure 15.1 *Various approximation methods for a non-Gaussian distribution: (a) assumed true distribution, (b) normal approximation, (c) piecewise linear function approximation, (d) step-function approximation, (e) Gaussian mixture approximation and (f) particle approximations.*

uniquely determined as $w = G(y,x)$, where G is a differentiable function with respect to y_0 (Kitagawa (1996)).

As examples of nonlinear functions $H(x,w)$ that satisfy the above-mentioned requirements, we can consider $H(x,w) = H_1(x) + w$ and $H(x,w) = e^x w$, and in these cases $G(y,x)$ are given by $G(y,x) = y - H_1(x)$ and $G(y,x) = e^{-x}y$, respectively. For simplicity, y_n and w_n are assumed to be one-dimensional models here. However, we can easily consider extensions to the multi-dimensional models.

In this chapter, we consider the problem of state estimation for the nonlinear non-Gaussian state-space model. As stated in the previous chapter, a set of observations obtained until time t is denoted as $Y_t \equiv \{y_1, \ldots, y_t\}$, and we consider a method of obtaining the distributions based on particle approximations of the predictive distribution $p(x_n | Y_{n-1})$, the filter distribution $p(x_n | Y_n)$ and the smoothing distribution $p(x_n | Y_N)$.

Figure 15.1 shows various methods of approximating a non-Gaussian distribution. Plot (a) shows the assumed true distribution, which has two peaks. Plot (b) shows the approximation by a single Gaussian distribution, which corresponds to an approximation by linear-Gaussian state-space model or an extended Kalman filter. It is obvious that if the true distribution has two or more peaks or the distribution is highly skewed, a good approximation cannot be obtained. Plots (c) and (d) show the approximation by a piecewise linear function with forty nodes, and by a step function with forty nodes, respectively. In the previous chapter, these approximations were applied to the non-Gaussian filter and the smoother. In practical use, however, we usually set the number of nodes to several hundred or more. Consequently, very precise approximations to any type of distribution can be obtained by these methods.

Plot (e) shows the Gaussian mixture approximations with five Gaussian components. The thin curves depict the contributions of Gaussian components and the bold curve shows the Gaussian-mixture approximation obtained by summing up these Gaussian components. The Gaussian-sum filter and the smoother can be derived by this approximation.

In the particle filter presented in this chapter, distinct from the approximations shown above, the true distribution is represented by using many particles. Each particle is considered as a realization from the true distribution. Plot (f) shows the location of 100 particles generated from the assumed true distribution shown in plot (a). Comparing plot (a) and plot (f), it can be seen that the density of the particles in plot (f) correspond to the height of the density function in plot (a).

Figure 15.2 compares the empirical distribution functions obtained from the particles and the true cumulative distribution functions. Plots (a) – (c) show the cases of $m = 10$, 100 and 1,000, respectively. The true distribution function and the empirical distribution function are illustrated with smooth curve and step function, respectively. For $m = 10$, the appearance of the empirical distribution function considerably differs from the true distribution function. However, as the number of particles increases to $m = 100$ and 1,000, we obtain closer approximations.

In the particle filter, the predictive distribution, the filter distribution and the smoothing distribution are approximated by m particles as shown in Table 15.1. The number of particles m is usually set between 1,000 and 100,000, and the actual number used is chosen depending on the complexity of the distribution and the model, and the required accuracy. The approximation of a density function by m particles is equivalent to approximating a cumulative distribution function by an empirical distribution function defined using m particles.

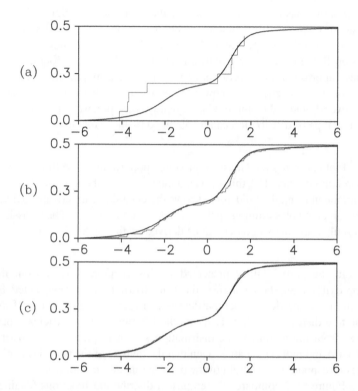

Figure 15.2 *Comparison between the empirical distribution functions and the true cumulative distribution functions for various numbers of particles: (a) m = 10, (b) m = 100 and (c) m = 1,000.*

Table 15.1 *Approximations of distributions used in the particle filter and the smoother.*

Distributions	Density functions	Approximations by particles
Predictive distribution	$p(x_n\|Y_{n-1})$	$\{p_n^{(1)},\ldots,p_n^{(m)}\}$
Filter distribution	$p(x_n\|Y_n)$	$\{f_n^{(1)},\ldots,f_n^{(m)}\}$
Smoothing distribution	$p(x_n\|Y_N)$	$\{s_{n\|N}^{(1)},\ldots,s_{n\|N}^{(m)}\}$
Distribution of system noise	$p(v_n)$	$\{v_n^{(1)},\ldots,v_n^{(m)}\}$

15.2 Particle Filter

The particle filter algorithm presented in this section is particularly useful to recursively generate the particles by approximating the one-step-
ahead predictive distribution and the filter distribution.

In particle filtering, the particles $\{p_n^{(1)}, \ldots, p_n^{(m)}\}$ that follow the predictive distribution, are generated from the particles $\{f_n^{(1)}, \ldots, f_{n-1}^{(m)}\}$
used for the approximation of the filter distribution of the previous state.
Then the realizations $\{f_n^{(1)}, \ldots, f_n^{(m)}\}$ of the filter can be generated by
re-sampling the realizations $\{p_n^{(1)}, \ldots, p_n^{(m)}\}$ of the predictive distribution (Gordon et al. (1993), Kitagawa (1996)).

15.2.1 One-step-ahead prediction

For the one-step-ahead prediction step, we assume that m particles
$\{f_{n-1}^{(1)}, \ldots, f_{n-1}^{(m)}\}$ can be considered as the realizations generated from
the filter distribution $p(x_{n-1}|Y_{n-1})$ of the previous step x_{n-1}, and the par-
ticles $v_n^{(1)}, \ldots, v_n^{(m)}$ can be considered as independent realizations of the
system noise v_n. That is, for $j = 1, \ldots, m$, it is assumed that

$$f_{n-1}^{(j)} \sim p(x_{n-1}|Y_{n-1}), \quad v_n^{(j)} \sim q(v). \tag{15.3}$$

Then, the particle $p_n^{(j)}$ defined by

$$p_n^{(j)} = F(f_{n-1}^{(j)}, v_n^{(j)}), \tag{15.4}$$

can be considered as a particle generated from the one-step-ahead predictive distribution of the state x_n. (See Appendix D.)

15.2.2 Filtering

In the next step of filtering, we compute $\alpha_n^{(j)}$, the Bayes importance factor (or likelihood) of the particle $p_n^{(j)}$ with respect to the observation y_n.
That is, for $j = 1, \ldots, m$,

$$\alpha_n^{(j)} = p(y_n|p_n^{(j)}) = r(G(y_n, p_n^{(j)})) \left| \frac{\partial G}{\partial y_n} \right|, \tag{15.5}$$

where G is the inverse function of H in the observation model, and r
is the probability density function of the observation noise w. Here $\alpha_n^{(j)}$

can be considered as a weighting factor representing the importance of the particle $p_n^{(j)}$.

Then, we obtain m particles $f_n^{(1)}, \ldots, f_n^{(m)}$ by re-sampling of $p_n^{(1)}, \ldots, p_n^{(m)}$ with sampling probabilities proportional to the "likelihoods" $\alpha_n^{(1)}, \ldots, \alpha_n^{(m)}$. Namely, obtain the sampling probability $\tilde{\alpha}_n^{(i)} = \alpha_n^{(i)} / (\alpha_n^{(1)} + \cdots + \alpha_n^{(m)})$ and obtain a new particle $f_n^{(j)}$ according to

$$
f_n^{(j)} = \left\{
\begin{array}{ccc}
p_n^{(1)} & \text{with probability } \tilde{\alpha}_n^{(1)} \\
\vdots & \vdots \\
p_n^{(m)} & \text{with probability } \tilde{\alpha}_n^{(m)}.
\end{array}
\right.
\tag{15.6}
$$

Then, $\{f_n^{(1)}, \ldots, f_n^{(m)}\}$ can be considered as the realizations generated from the filter distribution $p(x_n|Y_n)$. (See Appendix D.)

15.2.3 Algorithm for the particle filter

In summary, the following algorithm for the particle filter is obtained;

1. Generate k-dimensional random numbers $f_0^{(j)} \sim p_0(x)$ for $j = 1, \ldots, m$.

2. Repeat the following steps for $n = 1, \ldots, N$.

   (i) For $j = 1, \ldots, m$,
   - Generate ℓ-dimensional random numbers $v_n^{(j)} \sim q(v)$.
   - Obtain the new particle $p_n^{(j)} = F(f_{n-1}^{(j)}, v_n^{(j)})$.
   - Evaluate the importance factor $\alpha_n^{(j)} = r(G(y_n, p_n^{(j)})) \left| \frac{\partial G}{\partial y_n} \right|$.
   - Normalize the importance factor by $\tilde{\alpha}_n^{(j)} = \alpha_n^{(j)} / \sum_{i=1}^m \alpha_n^{(i)}$.

   (ii) Generate $\{f_n^{(1)}, \ldots, f_n^{(m)}\}$ by repeated re-sampling (sampling with replacement) m times from $\{p_n^{(1)}, \ldots, p_n^{(m)}\}$ with probabilities $\{\tilde{\alpha}_n^{(1)}, \ldots, \tilde{\alpha}_n^{(m)}\}$.

15.2.4 Likelihood of a model

The state-space models defined by (15.1) and (15.2) usually contain several unknown parameters $\theta = (\theta_1, \ldots, \theta_k)^T$, such as the variance of the noise and the coefficients of the nonlinear functions F and H. When the observations $y_1, \ldots, y_N$ are given from (9.21), the log-likelihood of the model specified by the parameter θ is given by

$$\ell(\theta) = \sum_{n=1}^{N} \log p(y_n|Y_{n-1}), \qquad (15.7)$$

where $p(y_1|Y_0)$ is taken as $p_0(y_1)$. Here, using the particle approximation of the predictive distribution,

$$p(y_n|Y_{n-1}) = \int p(y_n|x_n)p(x_n|Y_{n-1})dx_n$$

$$\cong \frac{1}{m}\sum_{j=1}^{m} p(y_n|p_n^{(j)}) = \frac{1}{m}\sum_{j=1}^{m} \alpha_n^{(j)}, \qquad (15.8)$$

the log-likelihood can be approximated by

$$\ell(\theta) = \sum_{n=1}^{N} \log p(y_n|Y_{n-1}) \cong \sum_{n=1}^{N} \log\left(\sum_{j=1}^{m} \alpha_n^{(j)}\right) - N\log m. \qquad (15.9)$$

The maximum likelihood estimate $\hat{\theta}$ of the parameter θ is obtained by numerically maximizing the above log-likelihood function. However, it should be noted that the log-likelihood obtained by the particle filter contains an inherent error due to the particle approximation, which consequently makes it difficult to obtain precise maximum likelihood estimates of the parameters. To avoid this problem, a self-organizing state-space model is proposed, in which the unknown parameter θ is included in the state vector (Kitagawa (1998)). With this method, we can estimate the state and the parameter simultaneously.

15.2.5 On the re-sampling method

Here we consider the *re-sampling method*, which is indispensable to the filtering step. The basic algorithm of the re-sampling method based on random sampling is as follows (Kitagawa (1996)):

For $j = 1, \ldots, m$, repeat the following steps (a)–(c).

(a) Generate uniform random number $u_n^{(j)} \in U[0, 1]$.

(b) Search for i that satisfies $\sum_{\ell=1}^{i-1} \tilde{\alpha}_n^{(\ell)} < u_n^{(j)} \le \sum_{\ell=1}^{i} \tilde{\alpha}_n^{(\ell)}$.

(c) Obtain a particle that approximates the filter distribution by
$$f_n^{(j)} = p_n^{(i)}.$$

It should be noted here that the objective of the resampling is to re-express the distribution function determined by the particles $\{p_n^{(1)}, \ldots, p_n^{(m)}\}$ with weights $\{\tilde{\alpha}_n^{(1)}, \ldots, \tilde{\alpha}_n^{(m)}\}$ by representing the empirical distribution function defined by re-sampled particles with equal weights. Accordingly, it is not essential to perform exact random sampling.

Considering the above, we can develop modifications of the re-sampling method with regard to sorting and random number generation. An important modification is the *stratified sampling*. In this method, divide the interval $[0, 1)$ into m sub-intervals, and obtain one particle $u_n^{(j)}$ from each sub-interval. In this case, the above step (a) is replaced with one of the following two steps:

(a-S)   Generate a uniform random number $u_n^{(j)} \sim U\left((j-1)/m, j/m\right]$,

(a-D)   For fixed $\alpha \in (0, 1]$, set $u_n^{(j)} = (j - \alpha)/m$.

Here, α may be an arbitrarily fixed real number $0 \le \alpha < 1$, e.g. $1/2$, or a uniform random number. From the results of the numerical experiment, it is seen that more accurate approximations were obtained by stratified re-sampling than the original random sampling.

15.2.6   Numerical examples

In order to exemplify how the one-step-ahead predictive distribution and the filter distribution are approximated by particles, we consider the result of one cycle of the particle filter, i.e., for $n = 1$. Assume that the following one-dimensional linear non-Gaussian state-space model is given;

$$\begin{aligned} x_n &= x_{n-1} + v_n, \\ y_n &= x_n + w_n, \end{aligned} \tag{15.10}$$

where v_n and w_n are white noises which follow the Cauchy distribution with probability density function $q(v) = \sqrt{0.1}\pi^{-1}(v^2 + 0.1)^{-1}$, and the standard normal distribution $N(0, 1)$, respectively. The initial state distribution $p_0(x_0)$ is assumed to follow the standard normal distribution. However, it is evident that normality of the distribution is not essential for the particle filtering method.

Under the above assumptions, the one-step-ahead predictive distribution $p(x_1|Y_0)$ and the filter $p(x_1|Y_1)$ were obtained. Here a small number of particles, $m = 100$, was used, in order to make features of the results clearly visible. In actual computations, however, we would use a larger number of particles for approximation.

The curve in Figure 15.3(a) shows the assumed distribution of the initial state $p_0(x_0)$. The vertical lines show the locations of 100 realizations generated from $p_0(x_0)$, and the histogram, which was obtained from these particles, approximates the true probability density function shown by bold curve. The bold curve in plot (b) shows the true distribution function of the initial state and the thin curve shows the empirical distribution function obtained from the particles shown in plot (a). Similar to these plots, plots (c) and (d) show the probability density function, the realizations and the cumulative distribution, together with its empirical counterpart of the system noise.

Plots (e) and (f) illustrate the predictive distribution, $p(x_1|Y_0)$. The curve in plot (e) shows the "true" probability density function obtained by numerical integration of the two density functions of plots (a) and (c). Plot (f) shows the "true" cumulative distribution function obtained by integrating the density function shown in plot (e). On the other hand, the vertical lines in plot (e) indicate the location of the particles $p_1^{(j)}$, $j = 1, \ldots, m$, obtained by substituting a pair of particles shown in plots (a) and (c) into the equation (D.4). The histogram defined by the particles $p_1^{(1)}, \ldots, p_1^{(m)}$ approximates the true density function shown in plot (e). The empirical distribution function and the true distribution function are shown in plot (f).

The curve in plot (g) shows the filter density function obtained from the non-Gaussian filter using the equation (14.5), when the observation $y_1 = 2$ is given. With respect to plot (g), the particles are located in the same place as plot (e); however, the heights of the lines are proportional to the normalized importance weight of the particle $\tilde{\alpha}_n^{(j)}$. Different from plot (f) for which the step sizes are identical, the cumulative distribution function in plot (h) approximates the filter distribution, although the locations of the steps of plot (h) are identical to those of plot (f). Plot (i) shows the locations of the particles, the histogram and the exact filter distribution after re-sampling. It can be seen that the density function and the cumulative distribution function in plots (i) and (j) are good approximations to those in plots (g) and (h), respectively.

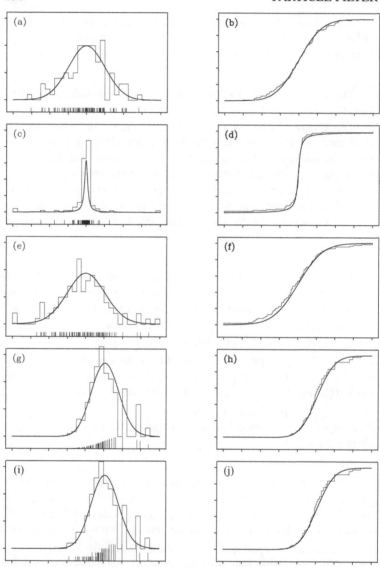

Figure 15.3 *One step of the particle filter. The figures in the left-hand column illustrate the probability density functions, 100 realizations and the histogram, respectively, and the figures in the right-hand column depict the distribution functions and the empirical distribution functions obtained from the realizations, respectively. (a) and (b): the initial state distributions. (c) and (d): the system noise distribution. (e) and (f): the one-step-ahead predictive distributions. (g) and (h): the filter distributions. (i) and (j): the filter distributions after resampling.*

15.3 Particle Smoothing Method

The particle filter method presented in the previous section, can be extended to a smoothing algorithm by preserving past particles. In the following, the vector of the particles $(s_{1|n}^{(j)}, \ldots, s_{n|n}^{(j)})^T$ denotes the j-th realization from the n-dimensional joint distribution function $p(x_1, \ldots, x_n | Y_n)$.

To achieve the smoothing, it is only necessary to modify Step 2(ii) of the algorithm discussed in Section 15.2 as follows:

(ii-S) For $i = 1, \ldots, m$, generate $(s_{1|n}^{(j)}, \ldots, s_{n-1|n}^{(j)}, s_{n|n}^{(j)})^T$, by re-sampling of the n-dimensional vector $(s_{1|n-1}^{(j)}, \ldots, s_{n-1|n-1}^{(j)}, p_n^{(j)})^T$.

In this modification, by re-sampling $\{(s_{1|n-1}^{j}, \ldots, s_{n-1|n-1}^{(j)}, p_n^{(j)})^T, j = 1, \ldots, m\}$ with the same resampling weights as used in Step (2)(ii) of Section 15.2.3, fixed-interval smoothing for a nonlinear non-Gaussian state-space model can be achieved (Kitagawa (1996)).

In actual computation, however, since a finite number of particles (m particles) is repeatedly re-sampled, the number of different particles gradually decreases, and the weights eventually become concentrated on a small number of particles, i.e. the distribution of the state collapses. Consequently, to obtain a practical smoothing algorithm, a modification for Step (ii-S) should be carried out as follows:

(ii-L) For $j = 1, \ldots, m$, generate $(s_{n-L|n}^{(j)}, \ldots, s_{n-1|n}^{(j)}, s_{n|n}^{(j)})$. Here, L is assumed to be a fixed integer, usually less than 30, and $f_n^{(j)} = s_{n|n}^{(j)}$ by re-sampling $(s_{n-L|n-1}^{(j)}, \ldots, s_{n-1|n-1}^{(j)}, p_n^{(j)})$.

It is interesting that this modified algorithm turns out to correspond to the L-lag fixed-lag smoothing algorithm. If L is set to a large value, the fixed-lag smoother $p(x_n | Y_{n+L})$ precisely approximates the fixed-interval smoother $p(x_n | Y_N)$. On the other hand, the distribution determined by the particles $x_{n|n+L}^{(1)}, \ldots, x_{n|n+L}^{(m)}$ may become far from $p(x_n | Y_{n+L})$. Therefore, L should be taken not so large, for example, $L = 20$ or 30.

The following example shows the results of state estimation by the particle filter and smoother for the artificially generated time series shown in Figure 14.1. The estimation is carried out by applying a first order trend model;

$$x_n = x_{n-1} + v_n,$$
$$y_n = x_n + w_n, \qquad (15.11)$$

where w_n is a Gaussian white noise with mean 0 and variance σ^2. For the system noise, two models are considered, namely, a Gaussian distribution and a Cauchy distribution.

The function pfilter of the TSSS package performs particle filtering and smoothing for a linear state-space model with non-Gaussian system noise and Gaussian observation noise. The function requires the following inputs:

y: a univariate time series.
m: number of particles.
model: assumed model for the system noise:
          = 0: normal distribution.
          = 1: Cauhy distribution.
          = 2: Gaussian mixture distribution.
lag: lag length in fixed-lag smoothing.
initd: initial distribution of the state:
          = 0: N(0,1).
          = 1: U(-4,4).
          = 2: Cauchy(0,1).
          = 3: delta function with mass at x=0.
sigma2:   observation noise variance σ^2.
tau2:   system noise variance τ^2.
plot: logical. If TRUE (default), posterior distribution is plotted.

and the outputs from this function are:
post.dist: posterior distributions.
llkhood: log-likelihood of the model.

Figure 15.4 (a) shows the exact filter distribution obtained by the Kalman filter when the system noise is Gaussian. The bold curve in the middle shows the mean of the distribution and the three gray areas illustrate $\pm 1\sigma$, $\pm 2\sigma$ and $\pm 3\sigma$ confidence intervals.

On the other hand, plots (b), (c) and (d) show the fixed-lag ($L = 20$) smoothed densities obtained by using $m = 1000, 10,000$ and $100,000$ particles, respectively. The dark gray area shows the $\pm 1\sigma$ interval and the light gray area shows the $\pm 2\sigma$ intervals. In plots (c) and (d), the $\pm 3\sigma$ interval is also shown.

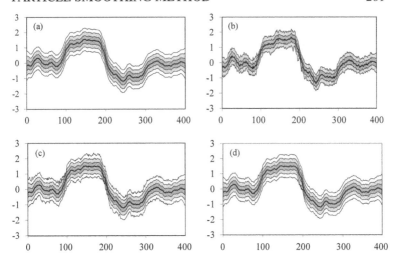

Figure 15.4 *The results of the particle filter: (a) the exact filter distribution using a Kalman filter.   (b) – (d) the fixed-lag* $(L = 20)$ *smoothed densities ((b)* $m =$ *1,000, (c)* $m = 10,000$, *(d)* $m = 100,000$*) with a particle filter.*

```
> # Input test data: level shift at N=101, 201 and 301
> y <- data( pfilter_sample )
>
> # Kalman filter
> # trend( y, trend.order = 1, tau2.ini = 1.4e-2, delta = 0.0 )
>
> # particle filter
> # system noise: Gaussian, observation noise: Gaussian
> # m = 1,000:
> pfilter(y, m = 1000 , model = 0, lag = 20 , initd = 0, sigma2
= 1.048, tau2 = 1.40e-2)
>
> # m = 10,000:
> pfilter(y, m = 10000, model = 1, lag = 20 , initd = 0, sigma2
= 1.048, tau2 = 1.40e-2)
>
> # m = 100,000:
> pfilter(y, m = 1000000, model = 2, lag = 20 , initd = 0,
sigma2 = 1.048, tau2 = 1.40e-2)
```

As can be seen in plot (b), there is a bias in the 50% percentile points and the 3σ interval cannot be drawn, but the general trend of the posterior distribution is captured with $m = 1000$. Plots (c) and (d) show that by

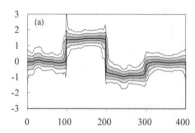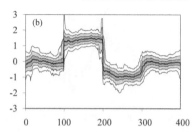

Figure 15.5 *Smoothing with Cauchy distribution model: (a) The exact distribution obtained from the non-Gaussian smoothing algorithm. (b) The results of particle smoothing (m = 100,000).*

increasing the number of particles, we can get a more accurate approximation of the "true" distribution. In particular, for $m = 100,000$ shown in plot (d), a very good approximation of the seven curves are obtained.

```
> # non-Gaussian smoother
> # system noise: Cauchy distribution
> ngsmth( y, noisev = 2, bv = 1.0, tau2 = 3.53e-5, noisew = 1,
sigma2 = 1.045 )
> #
> # particle filter and smoother
> # system noise: Cauchy distribution
> pfilter( y, m = 100000, model = 1, lag = 20 , initd = 0,
sigma2 = 1.045, tau2 = 3.53e-5)
```

Figure 15.5 shows the results when the system noise density is assumed to be a Cauchy distribution. The figure on the left-hand side shows the "exact" results obtained using the non-Gaussian smoothing algorithm presented in Chapter 14. On the other hand, the figure on the right-hand side depicts the results obtained by particle smoothing with $m = 100,000$ and $L = 20$. Although the curves are more variable in comparison with the figure on the left-hand side, the abrupt changes around $n = 100, 200$ and 300 are captured in a reasonable way. Moreover, the $\pm 3\sigma$ interval is also well approximated.

15.4 Nonlinear Smoothing

The particle filtering and smoothing algorithms can also be used for state estimation of the nonlinear state-space model (Kitagawa (1991)). Figures 15.6 (a) and (b) show examples of series x_n and y_n generated by the nonlinear state-space model

$$x_n = \frac{1}{2}x_{n-1} + \frac{25x_{n-1}}{x_{n-1}^2 + 1} + 8\cos(1.2n) + v_n,$$

$$y_n = \frac{x_n^2}{10} + w_n, \tag{15.12}$$

where $v_n \sim N(0,1)$, $w_n \sim N(0,10)$, $v_0 \sim N(0,5)$. Here, we consider the problem of estimating the unknown state x_n based on 100 observations, y_n, $n = 1,\ldots,100$. Because of the nonlinearity and the sinusoidal input in the system model in equation (15.12), the state x_n occasionally shifts between the positive and negative regions. However, since in the observation model, the state x_n is squared and contaminated with an observation noise, it is quite difficult to discriminate whether the true state is positive or negative.

The function `pfilterNL` of the TSSS package performs particle filtering and smoothing for a nonlinear state-space model. The function requires the following inputs:

y:	a univariate time series.
m:	number of particles.
lag:	lag length in fixed-lag smoothing.
sigma2:	observation noise variance σ^2.
tau2:	system noise variance τ^2.
plot:	logical. If TRUE (default), posterior distribution is plotted.

and the outputs from this function are:

post.dist:	posterior distributions.
	post.dist[,4]: median, post.dist[,j]: j=1,2,3,5,6,7
llkhood:	log-likelihood of the model.

```
> data(NLmodel)
> x <- NLmodel[, 1]
> pfilterNL(x, m = 100000, lag = 20 , sigma2 = 10.0, tau2 =
1.0, xrange = c(-20, 20), seed = 2019071117)
```

It is known that the extended Kalman filter occasionally diverges for this model (Kitagawa (1991)). Figure 15.6 (c) shows the smoothed posterior distribution $p(x_n|Y_N)$ obtained by the particle filter and smoother. It can be seen that a quite reasonable estimate of the state x_n is obtained with this method.

The key to successful non-linear filtering and smoothing is to allow the particle filter to reasonably represent non-Gaussian density of states.

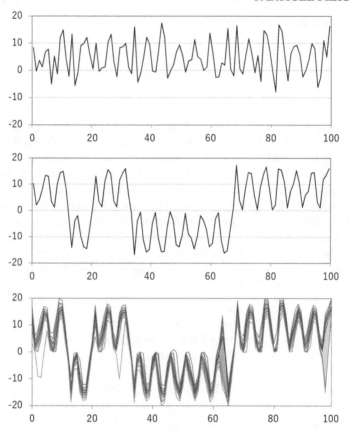

Figure 15.6 *Nonlinear smoothing: (a) data* y_n. *(b) unknown state* x_n. *(c) smoothed distribution of* x_n *obtained by the particle smoother.*

Figure 15.7 (a) and (b), respectively, show the marginal posterior state distributions $p(x_t|y_{1:t+j})$, $j = -1, 0, 1, 2$, for $t = 30$ and 48. For both $t = 30$ and 48, the predictive distribution $p(x_t|y_{1:t-1})$ and the filter distribution $p(x_t|y_{1:t})$ are bimodal. However, in the two-lag smoother for $t = 30$, $p(x_{30}|y_{1:32})$, the left-half of the distribution disappeared and the distribution becomes unimodal. The same phenomenon can be seen for $t = 48$. In this case, the right peak is higher than the left one in the predictive distribution. However, in the one-lag and two-lag smoother distributions, the peak in the right half domain disappears. In such a situation, the extended Kalman filter is very likely to yield an estimate with a reverse sign.

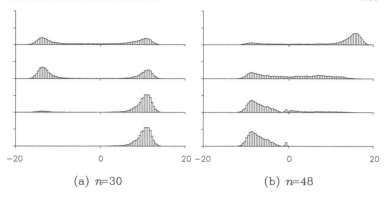

(a) n=30   (b) n=48

Figure 15.7 *Fixed-point smoothing for t = 30 (left) and t = 48 (right). From top to bottom, predictive distributions, filter distributions, one-lag smoothers and two-lag smoothers.*

Problems

1. Consider a model for the Nikkei 225 data shown in Figure 1(g) that takes into account changes in the trend and the volatility simultaneously.

2. In particle filtering, how many particles should we use to increase the accuracy of the estimate 10-fold.

3. For the particle filter, if we do not restrict every particle to have the same weight but allow some to have different weights, is it possible to develop a procedure without re-sampling?

4. Compare the amount of information required to be stored to carry out fixed-interval smoothing and fixed-lag smoothing.

Chapter 16

Simulation

In this chapter, we first explain methods for generating random numbers that follow various distributions. A realization of white noise can be obtained by repeating the generation of random numbers that follow a specified distribution. In time series modeling, a time series is considered as an output of a system with a white noise input. Therefore, if a time series model is given, we can generate a time series by using random numbers. In this chapter, a unified method for simulating time series is presented using the state-space model. Readers not interested in the details of random number generation may skip the first two sections.

16.1 Generation of Uniform Random Numbers

A sequence of independently generated numbers that follows a certain distribution is called a sequence of random numbers or simply *random numbers*. In other words, the random numbers are realizations of white noises. A time series model represents a time series as the output of a system that has white noise as an input. Therefore, we can generate a time series by properly specifying the time series model and the random numbers, and this method is called a *simulation* of a time series model.

In actual computation for simulating a time series, we usually use pseudo-random numbers generated using an appropriate algorithm. Normal random numbers, i.e. random numbers which follow a normal distribution, are frequently necessary for simulating time series. Such a sequence of random numbers is obtained by generating *uniform random numbers*, and then transforming them to the specified distribution.

A well-known conventional method for generating uniform random numbers is the *linear congruence method*. With this method, a sequence of integers $I_1, I_2, \cdots$ is generated from

$$I_k \equiv aI_{k-1} + c \quad (\text{mod } m), \tag{16.1}$$

starting from an initial integer I_0.

In particular, for $c = 0$, this method is called the *multiplicative congruence method*. The generated integer I_k takes a value in $[0, m)$. Therefore, I_k/m is distributed on $[0, 1)$.

However, we note here that since the period of the series generated by this method is at most m, it is not possible to generate a sequence of random numbers longer than m. The constants a, c and m have to be selected carefully, and examples of the combination of constants are:

$$a = 1229 \qquad c = 351750 \qquad m = 1664501$$
$$a = 1103515245 \qquad c = 12345 \qquad m = 2^{31}$$

(Knuth (1997). The latter combination were used in C language until 1990-th. It is known that if $m = 2^p$, $a = 5 \pmod{m}$ and c is an odd number, then the period of the sequence generated by this algorithm attains the possible maximum period of m, i.e., it containes all of the number between 0 and $m - 1$ once each.

By the lagged Fibonacchi method, the sequence of integers I_n is generated by

$$I_{n+p} = I_{n+q} + I_n \pmod{m}, \tag{16.2}$$

where m is usually set to 2^k (k is the bit length). Then the generated series has the period at largest $2^{k-1}(2^p - 1)$. In the current C language, $p = 31$, $q = 3$, $m = 2^{31}$ are used that attain the period of about 2^{60}.

In the generalized feedback shift register (GFSR) algorithm (Lewis and Payne (1973)), k-dimensional vector of bianary sequence is generated by

$$I_{n+p} = I_{n+q} \oplus I_n, \tag{16.3}$$

where $p > q$ and $\oplus$ denotes the excusive OR operation. The integers p and q are determined so that the polynomial $X^p + X^q + 1$ is irreducible. Then the period of the sequence becomes at largest $2^p - 1$. Some examples of p and q are; $p = 521$, $q = 158$ (period $2^{521} - 1$), $p = 521$, $q = 32$ (period $2^{521} - 1$) and $p = 607$, $q = 273$ (period $2^{607} - 1$).

A twisted GFSR algorithm (Matsumoto and Kurita (1994)) generates a k-dimensional bianary sequence by

$$I_{n+p} = I_{n+q} \oplus I_n A, \tag{16.4}$$

where A is a $k \times k$ regular matrix, such as the companion matrix. In the frequently used TT800 algorithm, $p = 25$ and $q = 7$ are used. This sequence has the period of $2^{pk} - 1$. Further, Matsumoto and Nishimura (1998) developped the Mersenne twister algorithm

$$I_{n+p} = I_{n+q} \oplus I_{n+1} B \oplus I_n A. \tag{16.5}$$

They showed that for $p = 624$, by taking A and B appropriately, the generated series has a period of length $2^{19937} - 1$ and are distributed uniformly on the 623-dimensional space.

Different from these families of pseudo-random numbers that are generated in software using certain algorithms, hardware for generating physical (hardware) random numbers has also been developed. Such hardware can be used when more precise random numbers are necessary in a simulation, because random numbers generated in this way are supposed to be free from any cycles or correlations.

Uniform random number generation in R

In the R environment the uniform random number can be obtained by the function `runif`, which is based on the Mersenne twister. In calling this function, the following inputs are required:

      n: size of random numbers.

      min, max: lower and upper limits of the distribution.

If `min` or `max` are not specified, they assume the default values of 0 and 1, respectively.

```
> # uniform random number generation
> y <- runif( 100 )
>
> y <- runif( 100, min = -1, max = 1)
```

16.2   Generation of White Noise

A realization of white noise is obtained by generating a sequence of random numbers that follows a specified distribution. In this section, we first consider the generation of Gaussian white noise, We then consider generation of white noise with other distributions such as the χ^2 distribution, Cauchy distribution and so on,

To generate Gaussian white noise, it is necessary to obtain a sequence of random numbers that follows a normal distribution, namely *normal random numbers*. The Box-Muller transformation (Box and Muller (1958)) for generating normal random numbers is well known. This method applies the fact that for two given independent uniform random numbers U_1 and U_2 on [0,1],

$$
\begin{aligned}
X_1 &= \sqrt{-2\log U_1}\cos 2\pi U_2, \\
X_2 &= \sqrt{-2\log U_1}\sin 2\pi U_2
\end{aligned}
\tag{16.6}
$$

independently follow the standard normal distribution $N(0, 1)$.

In practice, however, Marsaglia's algorithm that follows can avoid the explicit evaluation of sine and cosine functions, and thus can generate normal random numbers more efficiently.

1. Generate the uniform random numbers U_1 and U_2.
2. Put $V_1 = 2U_1 - 1$ and $V_2 = 2U_2 - 1$.
3. Put $S^2 = V_1^2 + V_2^2$.
4. Return to Step (1), if $S^2 \geq 1$.
5. Put $X_1 = V_1\sqrt{\frac{-2\log S}{S}}$ and $X_2 = V_2\sqrt{\frac{-2\log S}{S}}$.

Generation of Gaussian white noise in R

In the R environment, the normal random number can be obtained by the function rnorm. The required inputs are:

    n: size of random numbers.

    mean: mean of the distribution.

    sd: standard deviation of the distribution.

If mean or sd are not specified, the default values of 0 and 1 are assumed, respectively.

```
> # standard normal random number generation
> y <- rnorm( 100 )
>
> # normal random numbers with mean 0 and standard deviation 2.
> y <- rnorm( 100, mean = 1, sd = 2)
```

```
> # uniform rundom numbers
> plot( runif(200), type = "l", ylim = c(0,1) )
>
> # standard normal random numbers
> y <- rnorm( 200 )
> plot( y, type = "l", ylim = c(-3,3) )
>
> hist( y )
> acf( y )
```

In Figure 16.1, plot (a) shows 200 uniform random numbers obtained by the function runif. On the other hand, plot (b) shows normal random numbers (white noise) obtained by the function rnorm. Plots in the third row shows the histogram and the sample autocorrelation function of the normal random numbers. Plots in the bottom row shows the histogram and the sample autocorrelation function obtained from random numbers with 2000 samples. It can be seen from the sample autocorrelation function, the rundom numbers can be considered as uncorrelated series.

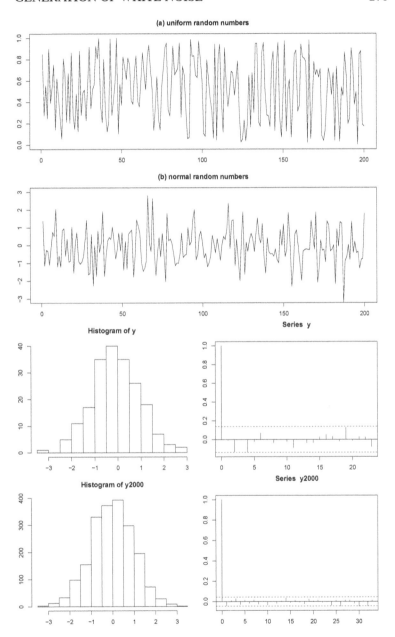

Figure 16.1 *Uniform random numbers, normal random numbers, histograms and autocorrelation functions.*

For some specific distributions such as the Cauchy distribution, the χ^2 distribution and the two-sided exponential (Laplace) distribution, they can be generated by a transformation of the uniform random numbers or the normal random numbers. In the following subsection, some examples are introduced.

16.2.1 χ^2 distribution

The χ^2 *distribution* with j degrees of freedom can be expressed as

$$\chi_j^2 = X_1^2 + \cdots + X_j^2, \tag{16.7}$$

where $X_1, \cdots, X_j$ are normal random variables. Therefore, to generate random numbers that follow the χ^2 distribution, we initially generate j normal random numbers and then obtain χ_j^2 using equation (16.7). In particular, *exponential random numbers* are obtained by putting $j = 2$. Further, by defining $Z = \log \chi_2^2$, we obtain the *double exponential random numbers*. Note that an exponential random number can be obtained from a uniform random number directly by setting

$$v = -\log u, \tag{16.8}$$

given the uniform random number u. Then the double exponential random number is efficiently obtained by

$$v = \log(-\log u). \tag{16.9}$$

16.2.2   Cauchy distribution

The density function of the *Cauchy distribution* that was used for the non-Gaussian model in Chapter 14 is given by

$$p(x) = \frac{1}{\pi(x^2 + 1)}. \tag{16.10}$$

To generate Cauchy random numbers, we generate a uniform random number on $[0, 1)$, and for $u \neq 0.5$, define v by

$$v = \tan \pi u. \tag{16.11}$$

16.2.3   Arbitrary distribution

In general, random numbers that follow an arbitrary density function $f(x)$ can be generated by using the inverse function of the distribution function. The distribution function is a monotone increasing function defined by

$$F(x) = \int_{-\infty}^{x} f(t)dt, \qquad (16.12)$$

which satisfies $0 \leq F(x) \leq 1$, $F(-\infty) = 0$, and $F(\infty) = 1$. Here, if we obtain an inverse function $G(y) \equiv F^{-1}(y)$, $0 \leq y \leq 1$, that satisfies $F(G(y)) = y$, then G is a function from $(0,1)$ to $(-\infty, \infty)$. Therefore, if u is a uniform random number on $[0,1)$, $v = G(u)$ becomes a random number that follows the density function $f(v)$. Applying this method, we can generate random numbers that follow various distributions by using the density functions discussed in Chapter 4.

In R environment, various random numbers can be generated by the function named rXXX. By changing xxx with a proper name, we can generate various continuous distributions such as:

rchisq: χ^2 distribution.
rt: t-distribution with df degree of freedom.
rcauchy: Cauchy distribution.
rexp: exponential ditribution

```
> # normal random number generation
> y <- rnorm( 100 )
>
> y <- rcauchy( 100, location = 0, scale = 1)
>
> y <- rt( 100, df = 3, ncp = 0 )
>
> y <- rchisq( 100, df = 2, ncp = 0 )
```

16.3   Simulation of ARMA models

Assume that the Gaussian white noise $v_1, \ldots, v_N$ is given. Then the simulation of the ARMA model

$$y_n = \sum_{j=1}^{m} a_j y_{n-j} + v_n - \sum_{j=1}^{\ell} b_j v_{n-j}, \qquad (16.13)$$

can be performed by using the initial values $y_1, \ldots, y_m$. If the initial values are not available, set $y_j = 0$ for $j = 1, \ldots, k$ where $k = \max(m, \ell)$. Then, for $n = k+1, \ldots, N$, it is possible to generate y_n by substituting $y_{n-1}, \ldots y_{n-m}$ and $v_n, v_{n-1}, \ldots, v_\ell$ into (16.13).

If we start the simulation from zero initial values, it is important to discard the first L data and use remaining $n = N - L$ data so that the effect of the initial condition disappears. For that purpose L needs to be selected sufficiently large so that the impulse response function defined in Chapter 6 can be considered as zero. More simply, take a sufficiently large value such as $L = 1000$.

On the other hand, if the past observations of the time series are available, we can use them as the initial value in simulation.

Example (Simulation of an AR model)

The following sample R code is to perform simulation by an AR model from the given initial values.

```
> n <- 200
> m <- 2
> ii <- m+1:n
> mm <- 1:m
> a <- c( 1.5,-0.8 )
> par( mfrow=c(2,1), mar=c(2,2,1,1) 0)
>
> # normal random number generation
> y <- rnorm( n )
> plot( y,type="l" )
>
> # generate AR process
> y[1] <- 0
> y[2] <- 0.5
>
> for( i in ii ){
> y[i] <- v[i]
> for( j in mm ) {
> y[i] <- y[i] + a[j]*y[i-j]
> }
> }
> plot( y,type="l" )
```

Figure 16.3 shows the results of the simulation by the second order AR model. Here the initial state is set to $y_1 = 0.0$, $y_2 = 0.5$ and generated 200 observations including 2 initial values using the Gaussian white noise v_n. The upper plot shows the white noise generated by the Gaussian white noise v_n. On the other hand, the bottom plot shows the generated AR process obtained by using the white noise.

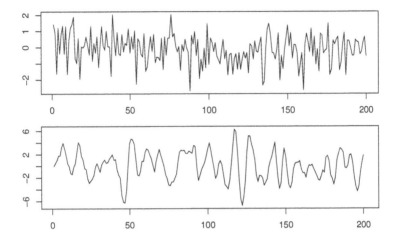

Figure 16.2: *Simulation of a Gaussian white noise and an AR model.*

16.4 Simulation Using a State-Space Model

In this section, we introduce a unified method for time series simulation using state space model. Assume that the time series model is represented in the form of the state-space model ;

$$x_n = Fx_{n-1} + Gv_n \tag{16.14}$$

$$y_n = Hx_n + w_n. \tag{16.15}$$

Based on this state-space model, if the initial state vector x_0, and N realizations of k-dimensional Gaussian white noise with mean 0 and variance covariance matrix Q, $v_1, \ldots, v_N$, are given, the realizations of the state vectors, $x_1, \ldots, x_N$ can be easily obtained by repeatedly substituting them into the system model of (16.14). Further, if the realizations of ℓ-dimensional Gaussian white noise with mean 0 and the variance covariance matrix R, $w_1, \ldots, w_N$, are given, the time series $y_1, \ldots, y_N$ can be generated by the observation model of (16.15).

A k-dimensional Gaussian white noise with mean 0 and variance covariance matrix Q can be obtained from k one-dimensional Gaussian white noises generated by the method shown in the previous subsection. Define a k-dimensional vector $u_n = (u_n^{(1)}, \cdots, u_n^{(k)})^T$, where $u_n^{(j)}$ is a one-dimensional Gaussian white noise, then u_n becomes a k-dimensional Gaussian white noise with mean vector 0 and unit variance covariance matrix I_k. Here we assume that a $k \times k$ lower triangular matrix L such that

$Q = LL^T$ can be obtained by the *Cholesky decomposition* of the positive-definite symmetric matrix Q. Then the k-dimensional vector v_n defined by

$$v_n = Lu_n \qquad (16.16)$$

satisfies

$$\mathrm{E}(v_n v_n^T) = \mathrm{E}(Lu_n u_n^T L^T) = L\mathrm{E}(u_n u_n^T)L^T = LI_k L^T = Q, \qquad (16.17)$$

showing that v_n becomes a normal white noise with mean vector 0 and the variance covariance matrix Q.

Simulation of Gaussian State-Space Model in R environment

```
> # Simulation by a state-space model
> simssm(n = nl, trend, seasonal.order = m2, seasonal, arcoef,
ar, tau1, tau2, tau3, sigma2 = z$sigma2, seed = 333)
```

The function `simssm` of the TSSS package performs simlation based on linear state-space model. The function requires the following inputs:

n:	number of simulated data.
trend.order:	order of trend component model (0 or1).
trend:	if trend.order>0, initial values of trend component:
seasonal.order:	order of seasonal component model (0,1,2).
seasonal:	if seasonal.order>0, initial values of seasonal component of length $p-1$:
p:	if seasonal.order>0, period of one season (typically 12).
ar.order:	order of AR component model (0 or positive integer).
ar:	if ar.order>0, initial values of AR component of length ar.order:
sigma2:	observation noise variance σ^2.
tau1:	if trend.order>0, variance of trend component model τ_1^2.
tau2:	if seasonal.order>0, variance of seasonal component model τ_2^2.
tau3:	if ar.order>0, variance of ar component model τ_3^2.
plot:	logical. If TRUE (default), simulated data are plotted.

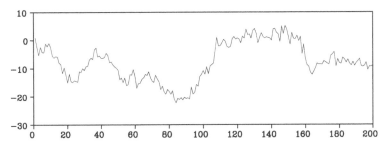

Figure 16.3: *Simulation of a random walk model.*

As the output from this function, an obect of class "simulate" gives simulated data of Gaussian state-space model.

```
> # Random walk model with obserrvation noise
> simssm(n = 200, seasonal.order = 0, arcoef=0.5, ar=0,
tau3=pi*pi/6, sigma2 = 1, seed = 333)
>
> # Second order AR model
> simssm(n = 200, seasonal.order = 0, arcoef=c(1.5,-0.8),
ar=c(0,0.1), tau3=1, sigma2 = 0, seed = 567)
```

Example (Simulation of a random walk model with observation noise) Figure 16.3 shows the results of the simulation of the random walk model with observation noise

$$
\begin{aligned}
x_n &= x_{n-1} + v_n, \\
y_n &= x_n + w_n,
\end{aligned}
\tag{16.18}
$$

where the system noise v_n follows the normal distribution $N(0, \pi^2/6)$ and the observation noise w_n is the standard Gaussian white noise that follows $N(0, 1)$ and the initial state is $x_0 = 0$.

Example (Simulation of a seasonal adjustment model)

The simulation of the seasonal adjustment model can be performed by the function simssm as follows:

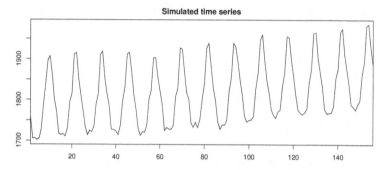

Figure 16.4: *Simulation of a seasonal adjustment model.*

```
> # Simulation of BLSALLFOOD data
> data(BLSALLFOOD)
> m1 <- 2; m2 <- 1; m3 <- 2
> z <- season(BLSALLFOOD, trend.order = m1, seasonal.order =
m2, ar.order = m3)
>
> nl <- length(BLSALLFOOD)
> trend <- z$trend[m1:1]
> arcoef <- z$arcoef
> period <- 12
> seasonal <- z$seasonal[(period-1):1]
> ar <- z$ar[m3:1]
> tau1 <- z$tau2[1]
> tau2 <- z$tau2[2]
> tau3 <- z$tau2[3]
> simssm(n = nl, trend, seasonal.order = m2, seasonal, arcoef,
ar, tau1, tau2, tau3, sigma2 = z$sigma2, seed = 333)
```

Figure 16.4 shows the results of a simulation for the seasonal adjustment model. The model is shown in Figure 12.6 with the stationary AR
component ($m_1 = 2$, $m_2 = 1$, $m_3 = 2$ and $p = 12$). As the initial state
vector x_0, the vector $x_{0|N}$ is used for the smoothed estimate of the ini-
tial vector which is obtained from the actual seasonal data. Although the
simulation of the seasonal adjustment model resembles the actual time
series shown in Figure 1.1(d), it is evident that the trend component of
Figure 16.4 is considerably different from that of Figure 1.1(d).

16.5   Simulation with the Non-Gaussian State-Space Model

The method of simulation for the state-space model presented in the previous section can be easily extended to simulation of a non-Gaussian
time series model. Consider a non-Gaussian state-space model,

$$x_n = Fx_{n-1} + Gv_n, \qquad (16.19)$$
$$y_n = Hx_n + w_n, \qquad (16.20)$$

where v_n and w_n are not necessarily Gaussian and are distributed according to the density functions $q(v)$ and $r(w)$, respectively.

For simplicity, it is assumed that the k components of v_n and the
ℓ components of w_n are mutually independent. To simulate the above
non-Gaussian state-space model, it is necessary to use a random number
generator that follow a specified non-Gaussian density function $p(x)$.

The function ngsim of the TSSS package performs simlation based
on linear non-Gaussian state-space model. The function requires the following inputs:

n:	number of simulated data.
trend.order:	order of trend component model (0 or1).
trend:	if trend.order> 0, initial values of the trend component:
seasonal.order:	order of seasonal component model (0,1,2).
seasonal:	if seasonal.order> 0, initial values of seasonal component of length $p-1$:
p:	if seasonal.order>0, period of one season (typically 12).
ar.order:	order of AR component model (0 or positive integer).
ar:	if ar.order>0, initial values of AR component of length ar.order.
noisew:	type of the observational noise.
	-1 : Cauchy random number.
	-2 : exponential distribution.
	-3 : double exponential distribution.
	0 : double exponential distribution (+ Euler's constant).
	1 : normal distribution,
	2 : Pearson distribution,
	3 : double exponential distribution.
wminmax:	lower and upper bound of observational noise.

paramw: parameter of the observational noise density.
            variance σ^2 if noisew = 1.
            dispersion parameter (σ^2) and shape parameter bw)
            if noisew = 2.
noisev: type of the system noise.
            -1 : Cauchy random number,
            -2 : exponential distribution,
            -3 : double exponential distribution,
            0 : double exponential distribution (+ Euler's constant),
            1 : normal distribution,
            2 : Pearson distribution,
            3 : double exponential distribution.
vminmax: lower and upper bound of system noise.
paramv: parameter of the system noise density.
            variance τ^2 if noisev = 1,
            dispersion parameter (τ^2) and shape parameter (bv) if
            noisev = 2.
plot: logical. If TRUE (default), simulated data are plotted.

As the output from this function, an object of class "simulate" gives
simulated data of Gaussian state-space model.

```
> # Simulation of the trend model with double exponential noise
inputs
> ar3 <- ngsim(n = 400, arcoef = 1.00, noisew = 1, paramw = 1,
noisev = 3, paramv = 1, seed = 555)
> plot(ar3, use = c(201, 400))
```

Example (Trend model with double-exponential noise)   Figure 16.5
shows the results of the simulation which were obtained by replacing the
system noise of the random walk model in Figure 16.2 by $v_n = r_n + \zeta$,
where r_n denotes a double exponential random number and $\zeta = 0.57722$
denotes the Euler constant.

For comparison, plots (a) and (b) show the density functions of the
normal distribution used in Figure 16.2 and the double exponential distribution. Even though the two density functions have the same mean and
the same variance, they have different shapes and so different simulation
results are obtained.

As shown in Figure 16.2, simulation with the density given in plot
(a) yields symmetric movement around the trend. On the other hand, in
plot (c) obtained by using the density in plot (b), the behavior around
the trend shows asymmetric upward and downward tendencies. We can

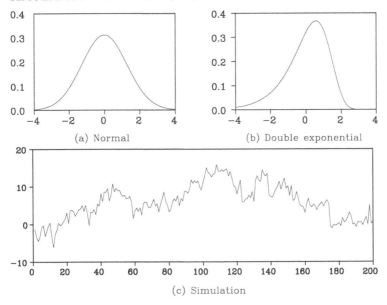

Figure 16.5: *Density functions of system noise and the results of the simulation.*

observe such typical asymmetric behavior of time series frequently in financial data.

Example (AR models with Gaussian and non-Gaussian noise)   Figure 16.6 shows the simulation results for the AR model of first order:

$$y_n = 0.9y_{n-1} + v_n. \qquad (16.21)$$

Here, in plots (a) and (b), v_n is assumed to be the standard normal distribution $N(0, 1)$ and the t-distribution, respectively. Even though the AR models are of the same order and have the same AR coefficients, the two time series in plots (a) and (b) appear quite different, because the distributions of the noise v_n are different. In particular, in plot (b), the time series occasionally shows big jumps, and the width of the fluctuation is, as a whole, more than twice as large as that of plot (a).

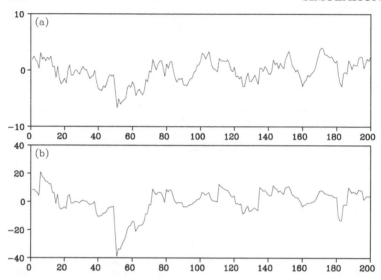

Figure 16.6 *Simulation of an AR model with different noise distributions: (a) normal distribution, (b) t-distribution.*

```
> ar1 <- ngsim(n = 400, arcoef = 0.90, noisew = 1, paramw = 1,
noisev = 1, paramv = 1, seed = 555)
> plot(ar1, use = c(201, 400))
>
> # Non-Gaussian (t-distribution) system noise
> ar1 <- ngsim(n = 400, arcoef = 0.95, noisew = 1, paramw = 1,
noisev = 2, paramv = c(1,1), seed = 555)
> plot(ar1, use = c(201, 400))
```

Problems

1. Assume that U and V are uniform random numbers defined on $[0, 1]$. What distribution does $W = U + V$ follow?

2. In the simulation of an AR model $y_n = ay_{n-1} + v_n$, $v_n \sim N(0, 1)$, we usually generate $m + n$ data elements and discard the initial m realizations. How large should m be?

3. Show a method of simulating future values based on observations up to the present time.

Appendix A

Algorithms for Nonlinear Optimization

In this appendix, we briefly discuss algorithms for nonlinear optimization. For simplicity, we consider the case where the objective function $f(x)$ is approximated by its Taylor series expansion up to the second order:

$$f(x) = f(x^*) + (x - x^*)^T g(x^*) + \frac{1}{2}(x - x^*)^T H(x - x^*), \qquad (A.1)$$

where $x = (x_1, \ldots, x_m)^T$ and

$$g(x) = \begin{bmatrix} \frac{\partial f(x)}{\partial x_1} \\ \vdots \\ \frac{\partial f(x)}{\partial x_m} \end{bmatrix}, \quad H = \begin{bmatrix} \frac{\partial^2 f}{\partial x_1 \partial x_1} & \cdots & \frac{\partial^2 f}{\partial x_1 \partial x_m} \\ \vdots & \ddots & \vdots \\ \frac{\partial^2 f}{\partial x_m \partial x_1} & \cdots & \frac{\partial^2 f}{\partial x_m \partial x_m} \end{bmatrix}. \qquad (A.2)$$

From (A.1), we have

$$g(x) = g(x^*) + H(x - x^*) \qquad (A.3)$$

and if x^* is assumed to minimize the objective function $f(x)$, $g(x^*) = 0$ holds and we have

$$g(x) = H(x - x^*). \qquad (A.4)$$

Therefore, if the current approximation is x, by computing H and $g(x)$, the solution x^* is obtained as

$$x^* = x - H^{-1}g(x). \qquad (A.5)$$

In actual problems, the objective function $f(x)$ is rarely expressible as a quadratic form such as that of (A.1), and the solution x^* is not obtained by applying (A.5) only once. Moreover, because the matrix H is not usually constant but changes depending on x, it is necessary to repeat the recursion

$$x_k = x_{k-1} - H_{k-1}^{-1} g(x_{k-1}), \qquad (A.6)$$

using an appropriately determined initial value x_0.

This procedure is called the *Newton-Raphson method*. However, since the likelihood function of the time series model is usually a very complex function of the parameters and it is sometimes difficult to obtain H_k analytically, the following algorithm, known as the *quasi-Newton method*, is used (Nocedal and Wright (2006)).

1. Put $k = 1$ and choose an initial value x_0 and H_0^{-1} suitably. As a default value, we may put $H_0 = I$.

2. Calculate the gradient vector $g(x_{k-1})$.

3. Obtain the direction vector by $h_k = H_{k-1}^{-1} g(x_{k-1})$.

4. Put $x_k = x_{k-1} - \lambda_k h_k$ and find λ_k, which minimizes $f(x_k)$ by a linear search.

5. Obtain an estimate of H_k^{-1} using either the DFP formula or the BFGS formula that follow.

[DFP formula] (Fletcher (1980))

$$H_k^{-1} = H_{k-1}^{-1} + \frac{\Delta x_k \Delta x_k^T}{\Delta x_k^T \Delta g_k} - \frac{H_{k-1}^{-1} \Delta g_k \Delta g_k^T H_{k-1}^{-1}}{\Delta g_k^T H_{k-1}^{-1} \Delta g_k}, \qquad (A.7)$$

[BFGS formula] (Broyden (1970))

$$H_k^{-1} = \left(I - \frac{\Delta g_k \Delta x_k^T}{\Delta g_k^T \Delta x_k} \right)^T H_{k-1}^{-1} \left(I - \frac{\Delta g_k \Delta x_k^T}{\Delta g_k^T \Delta x_k} \right) + \frac{\Delta x_k \Delta x_k^T}{\Delta x_k^T \Delta g_k}, \qquad (A.8)$$

where $\Delta x_k = x_k - x_{k-1}$ and $\Delta g_k = G(x_k) - g(x_{k-1})$.

6. If the convergence condition is not satisfied, go back to step 2 after setting $k = k + 1$.

This quasi-Newton method has the following two advantages. Firstly, it is not necessary to compute the Hessian matrix H_k and its inverse H_k^{-1} directly. Secondly, it can usually find the minimum even when the objective function is far from being a quadratic function, since it automatically adjusts the step width λ_k by a linear search. BFGS formula has an advantage that it does not require precise linear search.

Derivation of Levinson's Algorithm

In this appendix, we derive the Levinson's algorithm (Levinson (1947)) for estimating the AR coefficients for univariate time series model effectively. Whittle (1963) showed an extension of an algorithm for multivariate time series. However, since for multi-variate time series, the coefficients of the backward AR model are different from those of the forward AR model, the algorithm needs to treat twice as many number of recursions as the univariate case.

Assume that the autocovariance function C_k, $k = 0, 1, \cdots$ is given and that the Yule-Walker estimates of the AR model of order $m - 1$ have already been obtained. In the following, since we consider AR models with different orders, the AR coefficients, the prediction error and the prediction error variance of the AR model of order $m - 1$ are denoted by $\hat{a}_1^{m-1}, \ldots, \hat{a}_{m-1}^{m-1}$, v_n^{m-1} and $\hat{\sigma}_{m-1}^2$, respectively. Next, we consider a method of efficiently obtaining Yule-Walker estimates $\hat{a}_1^m, \ldots, \hat{a}_m^m$ and $\hat{\sigma}_m^2$ for the parameters of the AR model of order m using the estimates for the AR model of order $m - 1$.

Since the coefficients $\hat{a}_1^{m-1}, \ldots, \hat{a}_{m-1}^{m-1}$ satisfy the following Yule-Walker equation of order $m - 1$, we have that

$$C_k = \sum_{j=1}^{m-1} \hat{a}_j^{m-1} C_{j-k}, \qquad k = 1, \ldots, m - 1. \qquad (B.1)$$

This means that v_n^{m-1} satisfies

$$\mathrm{E}\left\{ v_n^{m-1} y_{n-k} \right\} = \mathrm{E}\left\{ \left(y_n - \sum_{j=1}^{m-1} \hat{a}_j^{m-1} y_{n-j} \right) y_{n-k} \right\} = 0, \qquad (B.2)$$

for $k = 1, \ldots, m - 1$. Here we consider a *backward AR model* of order $m - 1$ that represents the present value of the time series y_n with future values $y_{n+1}, \ldots, y_{n+m-1}$,

$$y_n = \sum_{i=1}^{m-1} a_j y_{n+i} + w_n^{m-1}. \qquad (B.3)$$

Then, since $C_{-k} = C_k$ holds for a univariate time series, this backward AR model also satisfies the same Yule-Walker equation (B.1). Therefore, the backward prediction error w_n^{m-1} satisfies

$$E\{w_{n-m}^{m-1}y_{n-k}\} = E\left\{\left(y_{n-m} - \sum_{j=1}^{m-1} \hat{a}_j^{m-1} y_{n-m+j}\right)y_{n-k}\right\} = 0, \quad \text{(B.4)}$$

for $k = 1,\ldots,m-1$. Here, we define the weighted sum of the forward prediction error v_n^{m-1} and the backward prediction error w_{n-m}^{m-1} by

$$
\begin{aligned}
z_n &\equiv v_n^{m-1} - \beta w_{n-m}^{m-1} \\
&\equiv y_n - \sum_{j=1}^{m-1} \hat{a}_j^{m-1} y_{n-j} - \beta\left(y_{n-m} - \sum_{j=1}^{m-1} \hat{a}_j^{m-1} y_{n-m+j}\right) \\
&= y_n - \sum_{j=1}^{m-1}\left(\hat{a}_j^{m-1} - \beta \hat{a}_{m-j}^{m-1}\right)y_{n-j} - \beta y_{n-m}.
\end{aligned}
\quad \text{(B.5)}
$$

z_n is a linear combination of $y_n, y_{n-1}, \ldots, y_{n-m}$ and from (B.2) and (B.4), it follows that

$$E\{z_n y_{n-k}\} = 0 \quad \text{(B.6)}$$

for $k = 1,\ldots,m-1$. Therefore, if the value of β is determined so that $E(z_n y_{n-m}) = 0$ holds for $k = m$, the Yule-Walker equation of order m will be satisfied. For that purpose, from

$$
\begin{aligned}
&E(z_n y_{n-m}) \\
&= E\left\{y_n - \sum_{j=1}^{m-1} \hat{a}_j^{m-1} y_{n-j} - \beta\left(y_{n-m} - \sum_{j=1}^{m-1} \hat{a}_j^{m-1} y_{n-m+j}\right)\right\}y_{n-m} \\
&= C_m - \sum_{j=1}^{m-1} \hat{a}_j^{m-1} C_{m-j} - \beta\left(C_0 - \sum_{j=1}^{m-1} \hat{a}_j^{m-1} C_j\right) = 0,
\end{aligned}
\quad \text{(B.7)}
$$

β is determined as

$$
\begin{aligned}
\beta &= \left(C_0 - \sum_{j=1}^{m-1} \hat{a}_j^{m-1} C_j\right)^{-1}\left(C_m - \sum_{j=1}^{m-1} \hat{a}_j^{m-1} C_{m-j}\right) \\
&= \left(\sigma_{m-1}^2\right)^{-1}\left(C_m - \sum_{j=1}^{m-1} \hat{a}_j^{m-1} C_{m-j}\right).
\end{aligned}
\quad \text{(B.8)}
$$

Therefore, if we denote this β as $\hat{a}_m^m$ and define a_j^m by

$$\hat{a}_j^m \equiv \hat{a}_j^{m-1} - \hat{a}_m^m \hat{a}_{m-j}^{m-1}, \quad \text{(B.9)}$$

then from (B.5) – (B.7), we have

$$E\left\{\left(y_n - \sum_{j=1}^{m} \hat{a}_j^m y_{n-j}\right)y_{n-k}\right\} = C_k - \sum_{j=1}^{m} \hat{a}_j^m C_{k-j} = 0 \tag{B.10}$$

for $k = 1, \ldots, m$.

In addition, the innovation variance of the AR model with order m can be expressed as

$$\begin{aligned}
\hat{\sigma}_m^2 &= C_0 - \sum_{j=1}^{m} \hat{a}_j^m C_j \\
&= C_0 - \sum_{j=1}^{m-1} \left(\hat{a}_j^{m-1} - \hat{a}_m^m \hat{a}_{m-j}^{m-1}\right)C_j - \hat{a}_m^m C_m \\
&= C_0 - \sum_{j=1}^{m-1} \hat{a}_j^{m-1} C_j - \hat{a}_m^m \left(C_m - \sum_{j=1}^{m-1} \hat{a}_{m-j}^{m-1} C_j\right) \\
&= \hat{\sigma}_{m-1}^2 - \hat{a}_m^m \left(C_m - \sum_{j=1}^{m-1} \hat{a}_{m-j}^{m-1} C_j\right). \tag{B.11}
\end{aligned}$$

Using (B.8), it follows that $C_m - \sum_{j=1}^{m-1} \hat{a}_{m-j}^{m-1} C_j = \hat{a}_m^m \hat{\sigma}_{m-1}^2$ and then $\hat{\sigma}_m^2$ is obtained as follows:

$$\hat{\sigma}_m^2 = \hat{\sigma}_{m-1}^2 \left(1 - (\hat{a}_m^m)^2\right). \tag{B.12}$$

Appendix C

Derivation of the Kalman Filter and Smoother Algorithms

In this appendix, a brief derivation of the Kalman filter (Kalman (1960)) and the fixed interval smoothing algorithm (Bierman (1977)) are given.

C.1 Kalman Filter

[One-step-ahead Prediction]

From $x_n = F_n x_{n-1} + G_n v_n$, we obtain

$$
\begin{aligned}
x_{n|n-1} &= \mathrm{E}\left(x_n | Y_{n-1}\right) \\
&= \mathrm{E}\left(F_n x_{n-1} + G_n v_n | Y_{n-1}\right) \\
&= F_n \mathrm{E}\left(x_{n-1} | Y_{n-1}\right) \\
&= F_n x_{n-1|n-1}, \quad\quad\quad\quad\quad\quad\quad\quad\quad\quad (C.1)
\end{aligned}
$$

$$
\begin{aligned}
V_{n|n-1} &= \mathrm{E}\left(x_n - x_{n|n-1}\right)^2 \\
&= \mathrm{E}\left(F_n(x_{n-1} - x_{n-1|n-1}) + G_n v_n\right)^2 \\
&= F_n \mathrm{E}\left(x_{n-1} - x_{n-1|n-1}\right)^2 F_n^T + G_n \mathrm{E}\left(v_n^2\right) G_n^T \\
&= F_n V_{n-1|n-1} F_n^T + G_n Q_n G_n^T, \quad\quad\quad\quad (C.2)
\end{aligned}
$$

where for simplicity of notation, $x x^T$ is abbreviated as x^2.

[Filter]

Denote the prediction error of y_n by ε_n; then from $y_n = H_n x_n + w_n$, it can be expressed as

$$
\begin{aligned}
\varepsilon_n &\equiv y_n - \mathrm{E}\left(y_n | Y_{n-1}\right) \\
&= H_n x_n + w_n - \mathrm{E}\left(H_n x_n + w_n | Y_{n-1}\right) \\
&= H_n x_n + w_n - H_n \mathrm{E}\left(x_n | Y_{n-1}\right) \\
&= H_n(x_n - x_{n|n-1}) + w_n. \quad\quad\quad\quad\quad\quad (C.3)
\end{aligned}
$$

Therefore, we have

$$
\mathrm{Var}(\varepsilon_n) = H_n V_{n|n-1} H_n^T + R_n, \quad\quad\quad\quad (C.4)
$$

289

$$
\begin{aligned}
\mathrm{Cov}(x_n, \varepsilon_n) &= \mathrm{Cov}(x_n, H_n(x_n - x_{n|n-1}) + w_n) \\
&= \mathrm{Var}(x_n - x_{n|n-1})H_n^T \\
&= V_{n|n-1}H_n^T.
\end{aligned}
\tag{C.5}
$$

Using the facts that $Y_n = \{Y_{n-1}, y_n\} = Y_{n-1} \oplus \varepsilon_n$ and that in the case of the normal distribution, the conditional expectation of x_n given Y_n is expressible by orthogonal projection, it follows that

$$
\begin{aligned}
x_{n|n} = \mathrm{E}(x_n|Y_n) &= \mathrm{Proj}(x_n|Y_n) \\
&= \mathrm{Proj}(x_n|Y_{n-1}, \varepsilon_n) \\
&= \mathrm{Proj}(x_n|Y_{n-1}) + \mathrm{Proj}(x_n|\varepsilon_n).
\end{aligned}
\tag{C.6}
$$

Because $\mathrm{Proj}(x_n|\varepsilon_n)$ is obtained by regressing x_n on ε_n, from (C.4) and (C.5),

$$
\begin{aligned}
\mathrm{Proj}(x_n|\varepsilon_n) &= \mathrm{Cov}(x_n, \varepsilon_n)\mathrm{Var}(\varepsilon_n)^{-1}\varepsilon_n \\
&= V_{n|n-1}H_n^T(H_nV_{n|n-1}H_n^T + R_n)^{-1}\varepsilon_n \\
&\equiv K_n\varepsilon_n.
\end{aligned}
\tag{C.7}
$$

Therefore, we have

$$
x_{n|n} = x_{n|n-1} + K_n\varepsilon_n.
\tag{C.8}
$$

In addition, from

$$
\begin{aligned}
V_{n|n-1} &= \mathrm{E}(x_n - x_{n|n-1})^2 \\
&= \mathrm{E}(x_n - x_{n|n} + K_n\varepsilon_n)^2 \\
&= V_{n|n} + K_n\mathrm{Var}(\varepsilon_n)K_n^T,
\end{aligned}
\tag{C.9}
$$

we have

$$
\begin{aligned}
V_{n|n} &= V_{n|n-1} - K_nH_nV_{n|n-1} \\
&= (I - K_nH_n)V_{n|n-1}.
\end{aligned}
\tag{C.10}
$$

C.2 Smoothing

$\delta_{n+1} \equiv x_{n+1} - x_{n+1|n}$ is assumed to be the prediction error of x_{n+1}. Define Z_n by

$$
Z_n \equiv Y_n \oplus \delta_{n+1} \oplus \{v_{n+1}, \ldots, v_N, w_{n+1}, \ldots, w_N\}.
$$

Then, we have the decomposition:

$$
\begin{aligned}
z_n &\equiv \mathrm{Proj}(x_n|Z_n) \\
&= \mathrm{Proj}(x_n|Y_n) + \mathrm{Proj}(x_n|\delta_{n+1}) \\
&\quad + \mathrm{Proj}(x_n|v_{n+1},\ldots,v_N,w_{n+1},\ldots,w_N) \quad\quad (C.11)
\end{aligned}
$$

and it follows that

$$
\begin{aligned}
\mathrm{Proj}(x_n|Y_n) &= x_{n|n} \\
\mathrm{Proj}(x_n|\delta_{n+1}) &= \mathrm{Cov}(x_n,\delta_{n+1})\mathrm{Var}(\delta_{n+1})^{-1}\delta_{n+1} \quad (C.12) \\
\mathrm{Proj}(x_n|v_{n+1},\ldots,v_N,w_{n+1},\ldots,w_N) &= 0.
\end{aligned}
$$

In addition, we have

$$
\begin{aligned}
\mathrm{Var}(\delta_{n+1}) &= V_{n+1|n}, \quad\quad (C.13) \\
\mathrm{Cov}(x_n,\delta_{n+1}) &= \mathrm{Cov}(x_n,F_{n+1}(x_n - x_{n|n}) + G_{n+1}v_{n+1}) \\
&= \mathrm{E}\,(x_n - x_{n|n})^2 F_{n+1}^T \\
&= V_{n|n}F_{n+1}^T. \quad\quad (C.14)
\end{aligned}
$$

Therefore, by putting $A_n = V_{n|n}F_{n+1}^T V_{n+1|n}^{-1}$, we have

$$
z_n = x_{n|n} + A_n(x_{n+1} - x_{n+1|n}). \quad\quad (C.15)
$$

Here, considering that Z_N generates Y_N, we obtain

$$
\begin{aligned}
x_{n|N} &= \mathrm{Proj}(x_n|Y_N) \\
&= \mathrm{Proj}(\mathrm{Proj}(x_n|Z_N)|Y_N) \\
&= \mathrm{Proj}(z_n|Y_N) \\
&= x_{n|n} + A_n(x_{n+1|N} - x_{n+1|n}). \quad\quad (C.16)
\end{aligned}
$$

Further, using

$$
x_n - x_{n|N} + A_n x_{n+1|N} = x_n - x_{n|n} + A_n x_{n+1|n}, \quad\quad (C.17)
$$

and $\mathrm{E}\left\{(x_n - x_{n|N})x_{n+1|N}^T\right\} = \mathrm{E}\left\{(x_n - x_{n|n})x_{n+1|n}^T\right\} = 0$, we obtain

$$
V_{n|N} + A_n \mathrm{E}\left\{x_{n+1|N}x_{n+1|N}^T\right\}A_n^T = V_{n|n} + A_n \mathrm{E}\left\{x_{n+1|n}x_{n+1|n}^T\right\}A_n^T. \quad (C.18)
$$

Here, using

$$
\mathrm{E}\left\{(x_{n+1} - x_{n+1|N})x_{n+1|N}^T\right\} = 0
$$
$$
\mathrm{E}\left\{(x_n - x_{n|n})x_{n+1|n}^T\right\} = 0,
$$

we obtain

$$
\begin{aligned}
E\left\{x_{n+1|N}x_{n+1|N}^{T}\right\} \\
&= E\left\{(x_{n+1|N}-x_{n+1}+x_{n+1})(x_{n+1|N}-x_{n+1}+x_{n+1})^{T}\right\} \\
&= V_{n+1|N}+E\left\{x_{n+1}x_{n+1}^{T}\right\}+2E\left\{(x_{n+1|N}-x_{n+1})x_{n+1}^{T}\right\} \\
&= V_{n+1|N}+E\left\{x_{n+1}x_{n+1}^{T}\right\}-2E\left\{(x_{n+1|N}-x_{n+1})(x_{n+1|N}-x_{n+1})^{T}\right\} \\
&= E\left\{x_{n+1}x_{n+1}^{T}\right\}-V_{n+1|N}, \qquad\qquad\qquad\qquad (C.19)
\end{aligned}
$$

and

$$
E\left\{x_{n+1|n}x_{n+1|n}^{T}\right\} = E\left\{x_{n+1}x_{n+1}^{T}\right\}-V_{n+1|n}. \qquad (C.20)
$$

Substituting this into (C.18) yields

$$
V_{n|N} = V_{n|n}+A_{n}(V_{n+1|N}-V_{n+1|n})A_{n}^{T}. \qquad (C.21)
$$

Appendix D

Algorithm for the Particle Filter

This appendix presents a derivation of the particle filter algorithm. The readers are referred to Kitagawa (1996) and Doucet et al. (2001) for details.

D.1 One-Step-Ahead Prediction

Assume that m particles $\{f_{n-1}^{(1)}, \ldots, f_{n-1}^{(m)}\}$ that can be considered as m independent realizations from the conditional distribution $p(x_{n-1}|Y_{n-1})$ of the state x_{n-1}, and m particles $\{v_n^{(1)}, \ldots, v_n^{(m)}\}$ from the system noise density $p(v)$ are given. Namely, assume that

$$f_{n-1}^{(i)} \sim p(x_{n-1}|Y_{n-1}), \quad v_n^{(i)} \sim q(v). \tag{D.1}$$

Then, it can be shown that the one-step-ahead predictive distribution of x_n is given by

$$
\begin{aligned}
p(x_n|Y_{n-1}) \\
= \int \int p(x_n, x_{n-1}, v_n|Y_{n-1})dx_{n-1}dv_n \\
= \int \int p(x_n|x_{n-1}, v_n, Y_{n-1})p(v_n|x_{n-1}, Y_{n-1})p(x_{n-1}|Y_{n-1})dx_{n-1}dv_n.
\end{aligned}
\tag{D.2}
$$

Since the system noise v_n is independent of past states and observations, the conditional distribution of the system noise satisfies $p(v_n|x_{n-1}, Y_{n-1}) = p(v_n)$. On the other hand, since x_n depends only on x_{n-1} and v_n, $p(x_n|x_{n-1}, v_n, Y_{n-1}) = p(x_n|x_{n-1}, v_n) = \delta(x_n - F(x_{n-1}, v_n))$, and we have

$$p(x_n|Y_{n-1}) = \int \int \delta(x_n - F(x_{n-1}, v_n))p(v_n)p(x_{n-1}|Y_{n-1})dx_{n-1}dv_n. \tag{D.3}$$

Therefore, when realizations $\{v_n^{(j)}\}$ of $p(v_n)$ and $\{f_{n-1}^{(j)}\}$ of $p(x_{n-1}|Y_{n-1})$ are given, realizations $\{p_n^{(j)}\}$ of $p(x_n|Y_{n-1})$ are obtained by

$$p_n^{(j)} = F(f_{n-1}^{(j)}, v_n^{(j)}). \tag{D.4}$$

293

D.2 Filter

When m independent realizations $p_n^{(1)}, \ldots, p_n^{(m)}$ of the distribution $p(x_n|Y_{n-1})$ are given, it is equivalent to approximate the distribution $p(x_n|Y_{n-1})$ by the empirical distribution function

$$P_n(x) = \frac{1}{m} \sum_{i=1}^{m} I(x, p_n^{(i)}), \tag{D.5}$$

where $I(x, a) = 0$ for $x < a$ and $I(x, a) = 1$ otherwise. This means that the predictive distribution $p(x_n|Y_{n-1})$ is approximated by the probability function

$$\Pr(x_n = p_n^{(j)}|Y_{n-1}) = \frac{1}{m}, \qquad \text{for } j = 1, \ldots, m. \tag{D.6}$$

Then, given the observation y_n, the posterior distribution of x_n is obtained by

$$
\begin{aligned}
\Pr(x_n = p_n^{(j)}|Y_n) &= \Pr(x_n = p_n^{(j)}|Y_{n-1}, y_n) \\
&= \lim_{\Delta y \to 0} \frac{\Pr(x_n = p_n^{(j)}, y_n \le y \le y_n + \Delta y|Y_{n-1})}{\Pr(y_n \le y \le y_n + \Delta y|Y_{n-1})} \\
&= \frac{p(y_n|p_n^{(j)})\Pr(x_n = p_n^{(j)}|Y_{n-1})}{\sum_{i=1}^{m} p(y_n|p_n^{(i)})\Pr(x_n = p_n^{(i)}|Y_{n-1})} \\
&= \frac{\alpha_n^{(j)} \cdot \frac{1}{m}}{\sum_{i=1}^{m} \alpha_n^{(i)} \cdot \frac{1}{m}} = \frac{\alpha_n^{(j)}}{\sum_{i=1}^{m} \alpha_n^{(i)}}.
\end{aligned}
\tag{D.7}
$$

The cumulative distribution function

$$\frac{1}{\sum_{i=1}^{m} \alpha_n^{(i)}} \sum_{i=1}^{m} \alpha_n^{(i)} I(x, p_n^{(i)}) \tag{D.8}$$

corresponding to this $\Pr(x_n = p_n^{(j)}|Y_n)$ has jumps with jump sizes proportional to $\alpha_n^{(1)}, \ldots, \alpha_n^{(m)}$ given by the right-hand side of the equation (15.9) only at the m points $p_n^{(1)}, \ldots, p_n^{(m)}$.

Although the approximation of the distribution of the filter was obtained using this expression (D.8), it is convenient to re-approximate it by m particles $f_n^{(1)},\ldots,f_n^{(m)}$ with equal weights to perform the computation of the prediction (D.1) in the next time step. This corresponds to representing the distribution of (D.8) by the empirical distribution function

$$\frac{1}{m}\sum_{i=1}^{m} I(x, f_n^{(i)}).$$ (D.9)

The m realizations $\{f_n^{(1)},\ldots,f_n^{(m)}\}$ can be obtained by re-sampling $\{p_n^{(1)},\ldots,p_n^{(m)}\}$ with probabilities

$$\Pr(f_n^{(j)} = p_n^{(i)}|Y_n) = \frac{\alpha_n^{(i)}}{\alpha_n^{(1)}+\cdots+\alpha_n^{(m)}}, \qquad j=1,\ldots,m.$$ (D.10)

D.3 Smoothing

For smoothing, assume that $\Pr(x_1 = s_{1|n-1}^{(j)},\ldots,x_{n-1} = s_{n-1|n-1}^{(j)}|Y_{n-1}) = 1/m$ and $v_n^{(j)} \sim q(v)$ and define $(p_{1|n-1}^{(j)},\ldots,p_{n|n-1}^{(j)})$ as follows,

$$p_{i|n-1}^{(j)} = \begin{cases} s_{i|n-1}^{(j)}, & \text{for } i = 1,\ldots,n-1 \\ F(s_{n-1|n-1}^{(j)}, v_n^{(j)}), & \text{for } i = n. \end{cases}$$ (D.11)

Then, $(p_{1|n-1}^{(j)},\ldots,p_{n|n-1}^{(j)})$ can be considered as a realization from the joint distribution of $(x_1,\ldots,x_n)$ when the observation Y_{n-1} is given. Next, given the observation y_n, the distribution $\Pr(x_1 \le p_{1|n-1}^{(j)},\ldots,x_n \le p_{n|n-1}^{(j)}|Y_{n-1})$ can be updated as follows;

$$\begin{aligned} &\Pr(x_1 = p_{1|n-1}^{(j)},\ldots,x_n = p_{n|n-1}^{(j)}|Y_n) \\ =\ & \Pr(x_1 = p_{1|n-1}^{(j)},\ldots,x_n = p_{n|n-1}^{(j)}|Y_{n-1},y_n) \\ =\ & \frac{p(y_n|x_1 = p_{1|n-1}^{(j)},\ldots,x_n = p_{n|n-1}^{(j)})}{p(y_n|Y_{n-1})} \\ & \times \Pr(x_1 = p_{1|n-1}^{(j)},\ldots,x_n = p_{n|n-1}^{(j)}|Y_{n-1}) \\ =\ & \frac{p(y_n|p_{n|n-1}^{(j)})\Pr(x_1 = p_{1|n-1}^{(j)},\ldots,x_n = p_{n|n-1}^{(j)}|Y_{n-1})}{p(y_n|Y_{n-1})}. \end{aligned}$$ (D.12)

Since $p_{n|n-1}^{(j)}$ is the same as the particle $p_n^{(j)}$ of the filter algorithm (D.5), the smoothing distribution $p(x_1,\dots,x_n|Y_n)$ can be obtained by re-sampling m n-dimensional vectors $(p_{1|n-1}^{(j)},\dots,p_{n|n-1}^{(j)})^T$, $j=1,\dots,m$ with the same weights as the filter.

Answers to the Problems

Chapter 1

1. It is important to select the sampling interval appropriately. If it is too wide, it is not possible to capture the features of the continuous time series. On the other hand, if it is too narrow, many parameters in the modeling may be used up on nonessentials, thus preventing effective representation of the time series. A rough rule of thumb is to set the sampling interval as $1/5$ to $1/10$ of the dominant period.

2. (An example.) Time series obtained by recording the sales amount of instant coffee at a store. Since values are non-negative, the distribution of the time series is asymmetric. Also this series may suddenly increase due to discounting or due to promotion activities.

3.(1) By solving (1.1) with respect to y, we obtain $y = e^z/(1+e^z)$.

(2) $z = \log\{(y-a)/(b-y)\}$, and the inverse transformation is given by $y = (a+be^z)/(1+e^z)$.

4. If the time series contains observation noise and is expressed as $y_n = a + bn + \varepsilon_n$, the difference of y_n becomes $\Delta y_n = b + \varepsilon_n - \varepsilon_{n-1}$. Therefore, although the trend component is removed, the noise components become more complex.

5. The corrected value of this year is affected by the change of the trend in the previous year. For example, if the slope of the trend increased in the middle of the previous year, the annual rate looks as if it decreased from the middle of this year.

6.(1) Assume that $y_n = T_n + w_n$, $T_n = a + bn$, and $w_n \sim N(0, \sigma^2)$ is a white noise, then

$$
\begin{aligned}
\hat{T}_n &= \frac{1}{3}(y_{n-1} + y_n + y_{n+1}) \\
&= \frac{1}{3}(T_{n-1} + T_n + T_{n+1}) + \frac{1}{3}(w_{n-1} + w_n + w_{n+1}).
\end{aligned}
$$

Here the first term on the right-hand side is T_n. On the other hand, the mean of the second term is 0 and from $E(w_{n-1}+w_n+w_{n+1})^2 = 3\sigma^2$, the variance becomes $\sigma^2/3$.

(2) By setting the number of terms large, we can get smoother estimates. On the other hand, this makes it difficult to detect sudden structural changes, and estimates may become sensitive to outlying observations. The moving median has the opposite properties.

Chapter 2

1. A Gaussian distribution is completely specified by its mean and variance. Therefore, if the mean and the variance are time-invariant, this also means that the distribution is time-invariant and becomes strongly stationary.

2. Consider a time series with a standard Cauchy distribution. Then obviously, it is strongly stationary. However, since the Cauchy distribution does not have a mean or variance, the series cannot be weakly stationary.

3. $C_k = E(y_n-\mu)(y_{n+k}-\mu) = E(y_{n+k}-\mu)(y_n-\mu) = E(y_n-\mu)(y_{n-k}-\mu) = C_{-k}$. The third equality holds only for a stationary time series y_n.

4. $C_0 = Ey_n^2 = Ev_n^2 - 2cEv_nv_{n-1} + c^2Ev_{n-1}^2 = 1 + c^2$, $C_1 = E(v_n - cv_{n-1})(v_{n-1} - cv_{n-2}) = -c$, $C_k = 0$ $(k \geq 2)$.

5. For arbitrary $\alpha = (\alpha_1,\ldots,\alpha_k)$ with not all components zero,

$$\alpha C\alpha^T = \sum_{i=1}^{k}\sum_{j=1}^{k}\alpha_i\alpha_jC_{i-j} = \sum_{i=1}^{k}\sum_{j=1}^{k}\alpha_i\alpha_jE(y_{n-i}y_{n-j})$$

$$= E\left(\sum_{i=1}^{k}\alpha_iy_{n-i}\right)^2 \geq 0.$$

6.(1) By taking the expectation of both sides,

$$E[\hat{C}_k(i,j)] = \frac{1}{N}\sum_{n=k+1}^{N}C_k(i,j) = \frac{N-k}{N}C_k(i,j).$$

(2) Since the sample autocovariance function defined by (2.21) is positive semi-definite, it has the significant advantage that the estimated AR models always become stationary. If N is replaced by $N - k$, we can get an unbiased estimate of $C_k(i - j)$. Instead however, we lose the advantage above of stationarity.

7.(1) When the sample size is n, $\hat{C}_0 \sim N(0, 2\sigma^4/n)$, $\hat{C}_k \sim N(0, \sigma^4/n)$, $\hat{R}_k \sim N(0, n^{-1})$.

(2) For all k, check if $|\hat{R}_k| < \frac{\alpha}{\sqrt{n}}$ holds. For example, α can be set equal to 2.

Chapter 3

1. From the expression $e^{-2\pi ikf} = \cos(2\pi kf) - i\sin(2\pi kf)$, we have
$p(f) = \sum\limits_{k=-\infty}^{\infty} C_k \cos(2\pi kf) - i \sum\limits_{k=-\infty}^{\infty} C_k \sin(2\pi kf)$. Here, since C_k is an even function and $\sin(2\pi kf)$ is an odd function, $C_k \sin(2\pi kf)$ becomes an odd function and the second term on the right-hand side of the above equation becomes 0.

2. $p(f) = 1 + C^2 - 2C\cos(2\pi f)$.

3.
$$
\begin{aligned}
p(f) &= \sum_{k=-\infty}^{\infty} C_k e^{-2\pi ikf} = \frac{\sigma^2}{1-a^2} \sum_{k=-\infty}^{\infty} a^{|k|} e^{-2\pi ikf} \\
&= \frac{\sigma^2}{1-a^2} \left(\frac{1}{1-ae^{-2\pi if}} + \frac{1}{1-ae^{2\pi if}} - 1 \right) \\
&= \frac{\sigma^2}{1-a^2} \frac{1-a^2}{|1-ae^{-2\pi if}|^2} = \frac{\sigma^2}{|1-ae^{-2\pi if}|^2}.
\end{aligned}
$$

4.
$$
\begin{aligned}
\frac{1}{N} &\left| \sum_{n=1}^{N} y_n \exp(-2\pi i(n-1)j/N) \right|^2 \\
&= \frac{1}{N} \sum_{n=1}^{N} \sum_{m=1}^{N} y_n y_m \exp(-2\pi i(n-1)j/N) \exp(2\pi i(m-1)j/N) \\
&= \frac{1}{N} \left\{ \sum_{n=1}^{N} y_n^2 + 2 \sum_{k=1}^{N-1} \sum_{n=k+1}^{N} y_n y_{n-k} \exp(-2\pi ikj/N) \right\} \\
&= \hat{C}_0 + 2 \sum_{k=1}^{N-1} \hat{C}_k \exp(-2\pi ikj/N).
\end{aligned}
$$

5. By taking the expectation of both sides of (3.9),
$$
\begin{aligned}
E[\hat{p}(f)] &= \sum_{k=1-N}^{N-1} E[\hat{C}_k] e^{-2\pi ikf} = \sum_{k=1-N}^{N-1} \frac{N-|k|}{N} C_k e^{-2\pi ikf} \\
&= \sum_{k=1-N}^{N-1} C_k e^{-2\pi ikf} - \frac{1}{N} \sum_{k=1-N}^{N} |k| C_k e^{-2\pi ikf},
\end{aligned}
$$

Chapter 4

1. The log-likelihood is given by

$$\ell = \sum_{i=1}^{n} \log f(m_i|\lambda) = \sum_{i=1}^{n}\{-\lambda + m_i \log \lambda - \log m_i!\}.$$

Therefore, from

$$\frac{\partial \ell}{\partial \lambda} = \sum_{i=1}^{n}\left\{-1 + \frac{m_i}{\lambda}\right\} = -n + \frac{1}{\lambda}\sum_{i=1}^{n} m_i = 0$$

we have $\hat{\lambda} = n^{-1}\sum_{i=1}^{n} m_i.$

2.(1)

$$\hat{\mu}_0 = \frac{1}{n+m}\left(\sum_{i=1}^{n} x_i + \sum_{i=1}^{m} y_i\right), \ \hat{\mu}_1 = \frac{1}{n}\sum_{i=1}^{n} x_i, \ \hat{\mu}_2 = \frac{1}{m}\sum_{i=1}^{m} y_i,$$

$$\hat{\sigma}_0^2 = \frac{1}{n+m}\left(\sum_{i=1}^{n}(x_i - \hat{\mu}_0)^2 + \sum_{i=1}^{m}(y_i - \hat{\mu}_0)^2\right),$$

$$\hat{\tau}_0^2 = \frac{1}{n+m}\left(\sum_{i=1}^{n}(x_i - \hat{\mu}_1)^2 + \sum_{i=1}^{m}(y_i - \hat{\mu}_2)^2\right).$$

Then by comparing

$$\text{AIC}_0 = (n+m)(\log 2\pi\hat{\sigma}_0^2 + 1) + 2 \times 2$$

$$\text{AIC}_1 = (n+m)(\log 2\pi\hat{\tau}_0^2 + 1) + 2 \times 3$$

and if $\text{AIC}_0 < \text{AIC}_1$, then this means that we can consider the two means to be identical.

(2) $\hat{\sigma}_1^2 = \frac{1}{n}\sum_{i=1}^{n}(x_i - \hat{\mu}_0)^2$, $\hat{\sigma}_2^2 = \frac{1}{m}\sum_{i=1}^{m}(y_i - \hat{\mu}_0)^2$, $\text{AIC}_2 = n(\log 2\pi\hat{\sigma}_1^2 + 1) + m(\log 2\pi\hat{\sigma}_2^2 + 1) + 2 \times 3$. If $\text{AIC}_0 < \text{AIC}_2$, then the two variances can be considered identical. Here from $\sum_{k=-\infty}^{\infty}|C_k| < \infty$, for arbitrary $\varepsilon > 0$, there exists an integer m that satisfies $\sum_{k=m+1}^{\infty}|C_k| < \varepsilon/2$. Then, we have

$$\left|\frac{1}{N}\sum_{k=-\infty}^{\infty}|k|C_k e^{-2\pi ikf}\right| \leq \frac{2}{N}\sum_{k=1}^{\infty}k|C_k| \leq \frac{2}{N}\sum_{k=1}^{m}kC_0 + 2\sum_{k=m+1}^{\infty}\frac{k}{N}|C_k|$$

$$\leq \frac{2}{N}\frac{m(m+1)}{2}C_0 + \varepsilon.$$

Since by taking N sufficiently large, the first term on the right-hand side of the equation can be made arbitrarily small, it follows that $E[\hat{p}(f)] \to p(f)$.

3. Putting $\hat{\mu} = \dfrac{1}{n}\sum_{i=1}^{n} y_i$, $\hat{\sigma}^2 = \dfrac{1}{n}\sum_{i=1}^{n}(y_i - \hat{\mu})^2$, the value of the AIC for the Gaussian distribution model is obtained by $\text{AIC}_0 = n\log 2\pi\hat{\sigma}^2 + n + 2 \times 2$. On the other hand, the AIC value for the Cauchy model is obtained by $\text{AIC}_1 = n\log \hat{\tau}^2 - 2n\log \pi - \sum_{i=1}^{n}\log\{(y_i - \hat{\mu})^2 + \hat{\tau}^2\} + 2 \times 2$, where $\hat{\mu}$ and $\hat{\tau}^2$ are obtained by maximizing (4.36). Then if $\text{AIC}_0 < \text{AIC}_1$, the Gaussian model is considered to be better.

4.(1) For the binomial distribution model $f(m|n, p) = {}_nC_m p^m (1 - p)^{n-m}$, the log-likelihood is defined by $\ell(p) = \log {}_nC_m + m\log p + (n - m)\log(1 - p)$. Therefore, from $\dfrac{d\ell}{dp} = 0$, $\dfrac{m}{p} - \dfrac{n-m}{1-p} = 0$, we obtain $\hat{p} = m/n$.

(2) The AIC value for the model (1) is obtained by $\text{AIC}_1 = -2\log {}_nC_m - 2m\log \hat{p} - 2(n - m)\log(1 - \hat{p}) + 2$. On the other hand, if we assume equal probabilities, taking $p = 1/2$, we obtain $\text{AIC}_0 = -2\log {}_nC_m + 2n\log 2$. Here if $\text{AIC}_0 < \text{AIC}_1$, we can consider that the equal probability model is better.

5. From $\dfrac{d}{d\theta}\log f(y|\theta) = f(y|\theta)^{-1}\dfrac{df(y|\theta)}{d\theta}$, we have

$$
\begin{aligned}
\frac{d^2}{d\theta^2}\log f(y|\theta) &= \frac{d}{d\theta}\left(\frac{d}{d\theta}\log f(y|\theta)\right) \\
&= \frac{d}{d\theta}\left(f(y|\theta)^{-1}\frac{df}{d\theta}\right) \\
&= -f(y|\theta)^{-2}\left(\frac{df}{d\theta}\right)^2 + f(y|\theta)^{-1}\frac{d^2f}{d\theta^2} \\
&= -\left(\frac{d}{d\theta}\log f(y|\theta)\right)^2 + f(y|\theta)^{-1}\frac{d^2f}{d\theta^2}.
\end{aligned}
$$

Therefore

$$
\begin{aligned}
J &= -E\left[\frac{d^2}{d\theta^2}\log f(y|\theta)\right] \\
&= E\left(\frac{d}{d\theta}\log f(y|\theta)\right)^2 - E\left[f(y|\theta)^{-1}\frac{d^2f}{d\theta^2}\right]
\end{aligned}
$$

$$= I - \int_{-\infty}^{\infty} \frac{d^2 f}{d\theta^2} dy = I - \frac{d^2}{d\theta^2} \int_{-\infty}^{\infty} f(y|\theta) dy = I.$$

6. From $z = k(y) = \log y$, $f(z) = (2\pi\sigma^2)^{-1} \exp\{-(z-\mu)^2/2\sigma^2\}$, and we have $\frac{dk}{dy} = y^{-1}$. Therefore

$$g(y) = \left|\frac{dk}{dy}\right| f(k(y)) = \frac{1}{y} \frac{1}{\sqrt{2\pi\sigma^2}} e^{-\frac{(\log y - \mu)^2}{2\sigma^2}}.$$

Chapter 5

1. Let $\hat{\mu}$ denote the maximum likelihood estimate of μ. Since the number of parameters is 1, the AIC is obtained by

$$\text{AIC} = n \log 2\pi\sigma^2 - \frac{1}{\sigma^2} \sum_{i=1}^{n} (y_i - \hat{\mu})^2 + 2.$$

2.(1) From $S = \sum_{n=1}^{N} (y_n - ax_n^2 - bx_n)^2$,

$$\frac{dS}{da} = -2 \sum_{n=1}^{N} x_n^2 (y_n - ax_n^2 - bx_n)$$

$$= -2 \left(\sum_{n=1}^{N} x_n^2 y_n - a \sum_{n=1}^{N} x_n^4 - b \sum_{n=1}^{N} x_n^3 \right) = 0$$

$$\frac{dS}{db} = -2 \sum_{n=1}^{N} x_n (y_n - ax_n^2 - bx_n)$$

$$= -2 \left(\sum_{n=1}^{N} x_n y_n - a \sum_{n=1}^{N} x_n^3 - b \sum_{n=1}^{N} x_n^2 \right) = 0.$$

Solving these, we have

$$\hat{a} = \frac{\sum x_n^2 y_n \sum x_n^2 - \sum x_n y_n \sum x_n^3}{\sum x_n^4 \sum x_n^2 - (\sum x_n^3)^2}$$

$$\hat{b} = \frac{\sum x_n y_n \sum x_n^4 - \sum x_n^2 y_n \sum x_n^3}{\sum x_n^4 \sum x_n^2 - (\sum x_n^3)^2}.$$

(2) $y_n = ax_n(x_n - b) + \varepsilon_n$. By minimizing $S = \sum_{n=1}^{N} (y_n - ax_n^2 - abx_n)^2$ with respect to the two parameters a and b by a numerical optimization procedure, we can obtain the least squares estimates.

Chapter 6

1.(1) Putting $m = 1$ in equation (6.21), the stationarity condition is that the root of $1 - a_1 B = 0$ lies outside the unit circle. Therefore, from $|B| = |a_1^{-1}| > 1$, we have $|a_1| < 1$.

(2) Putting $m = 2$ in equation (6.21), the roots are obtained by $B = (a_1 \pm \sqrt{a_1^2 + 4a_2})/2$. Therefore, the stationarity conditions are that $a_1^2 + 4a_2 \geq 0$, $a_2 < 1 - a_1$ and $a_2 < 1 + a_1$ for $a_1^2 + 4a_2 < 0$, $a_2 > -1$. Therefore, the region satisfying the stationarity condition is a triangle surrounded by three lines $a_2 < 1 - a_1$, $a_2 < 1 + a_1$ and $a_2 > -1$.

2.(1) σ^2

(2) From $y_{n+2} = a y_{n+1} + v_{n+2} = a^2 y_n + a v_{n+1} + v_{n+2}$, we have $y_{n+2|n} = a^2 y_n$. Therefore, $E\left(y_{n+2} - y_{n+2|n}\right)^2 = a^2 E v_{n+1}^2 + E v_{n+2}^2 = (1 + a^2)\sigma^2$.

(3) It can be expressed as $y_{n+k} = v_{n+k} + g_1 v_{n+k-1} + g_2 v_{n+k-2} + \cdots$. Here from the impulse response function of AR(1), $g_i = a^i$. Then, since we can express $y_{n+k|n} = g_k v_n + g_{k+1} v_{n+1} + \cdots$, we have

$$
\begin{aligned}
E\left(y_{n+k} - y_{n+k|n}\right)^2 &= E\left(v_{n+k} + g_1 v_{n+k-1} + \cdots + g_{k-1} v_{n+1}\right)^2 \\
&= (1 + g_1^2 + \cdots + g_{k-1}^2)\sigma^2 \\
&= (1 + a^2 + \cdots + a^{2(k-1)})\sigma^2 = \frac{1 - a^{2k}}{1 - a^2}\sigma^2.
\end{aligned}
$$

3.(1) From $C_0 = -0.9 C_1 + 1$, $C_1 = -0.9 C_0$, we have $C_0 = 0.81 C_0 + 1$, $C_0 = 1/0.19 = 5.26$, $C_k = 5.26 \times (-0.9)^k$.

(2) $C_0 = 25/7$, $C_1 = 75/28$, $C_3 = 15/14$, $C_4 = -9/28$.

(3) $C_0 = 1 + b^2$, $C_1 = -b$, $C_k = 0$ $(k > 1)$.

(4) Putting $E[y_n v_{n-k}] = g_k$, we have $g_0 = 1$, $g_1 = a - b$. Therefore, from $C_0 = a C_1 + g_0 - b g_1$, $C_1 = a C_0 - b g_0$, we have

$$
C_0 = \frac{1 - 2ab + b^2}{1 - a^2}, \qquad C_k = \frac{(1 - ab)(a - b)}{1 - a_2} a^{k-1}.
$$

4.(1) Substituting $v_n = y_n - a y_{n-1}$ into both sides of $v_n = b v_{n-1} + w_n$, yields

$$
y_n = (a + b) y_{n-1} - ab y_{n-2} + w_n,
$$

and we obtain an AR model of order 2.

(2) The autocovariance function of y_n is given by $a^k(1-a^2)^{-1}$, $k = 0, 1, 2, 3$. Therefore, $C_0 = (1-a^2)^{-1} + 0.01$, $C_k = a^k(1-a^2)^{-1}$.

5.(1) Substituting $C_0 = 1+b^2$, $C_1 = -b$, $C_k = 0$ and $k > 1$ into $p(f) = \sum_{k=-\infty}^{\infty} C_k e^{-2\pi ikf}$, we have $p(f) = (1+b^2) - be^{-2\pi if} - be^{2\pi if} = |1 - be^{-2\pi if}|^2$.

(2) From $dp/df = -4\pi a \sin(2\pi f)/\{1 - 2a\cos(2\pi f) + a^2\}^2 = 0$, we have $\sin(2\pi f) = 0$. This means that the maximum (or minimum) of the spectrum is attained either at $f = 0$ or $1/2$. If $a > 0$, it has a maximum at $f = 0$ and a minimum at $f = 1/2$. If $a < 0$, they are the other way around.

6. For $y_n = v_n - bv_{n-1} = (1-bB)v_n$, since $(1-bB)^{-1} = 1 + bB + b^2B^2 + \cdots$, we obtain an AR model of infinite order $y_n = -by_{n-1} - b^2y_{n-1} - \cdots + v_n$.

7.(1) From $y_{n+1|n} = -bv_n$ and $y_{n+k|n} = 0$, we have $E\varepsilon_{n+1|n}^2 = 1$, $E(\varepsilon_{n+k|n}^2) = 1 + b^2$ ($k \geq 1$).

(2) $y_n = v_n a v_{n-1} + a^2 v_{n-2} + \cdots$, $E(\varepsilon_{n+k|n}^2) = 1 + a^2 + \cdots + a^{2(k-1)}$.

(3) From $y_n = v_n + v_{n-1} + v_{n-2} + \cdots$, we have $E(\varepsilon_{n+k|n}^2) = k$.

Chapter 7

1.(1) $\text{AIC}_m = N(\log 2\pi\hat{\sigma}_{n-1}^2 + \log(1 - (a_m^m)^2) + 2(m+1) = \text{AIC}_{m-1} + N\log(1 - (a_m^m)^2) + 2$. Therefore, if $N\log(1 - (a_m^m)^2) + 2 < 0$, i.e., if $(a_m^m)^2 > 1 - e^{-2/N}$, then we can conclude that $\text{AR}(m)$ is better than $\text{AR}(m-1)$.

(2) $\sigma_1^2 = (1 - 0.9^2) \times 1 = 0.19$, $\sigma_2^2 = 0.1216$, $\sigma_3^2 = 0.1107$, $\sigma_4^2 = 0.1095$, $\sigma_5^2 = 0.1071$ (displayed up to the 4th decimal place).

(3) $\text{AIC}_0 = 100(\log 2\pi \times 1+) + 2 \times 1 = 285.78$
$\text{AIC}_1 = 100(\log 2\pi \times 0.19 + 1) + 2 \times 2 = 121.71$
$\text{AIC}_2 = 79.09$, $\text{AIC}_3 = 71.65$, $\text{AIC}_4 = 72.65$, $\text{AIC}_5 = 72.37$.
Therefore, the third order model is considered best.

2. The Yule-Walker method has the advantage that it always yields a stationary model. On the other hand, it may have large bias, especially for small sample sizes. These properties are due to the implicit assumption that the time series takes a value 0 outside the actually observed interval. The least squares method has the advantages that it provides us with estimates having a small bias and that it is easy to pool additional data and perform various manipulations for fitting

a more sophisticated model. On the other hand, it has problems in that stationarity is not guaranteed, and with parameter estimation, the first several observations are used only for conditioning and cannot be directly used for parameter estimation. In AR model estimation, the PARCOR method is very efficient in terms of computation and yields very close estimates as maximum likelihood estimates.

3. Fit a univariate AR model to each time series x_n and y_n, and obtain AIC_x and AIC_y, respectively. On the other hand, fit a two-variate AR model to $(x_n, y_n)^T$ and compute AIC_0. Then, if $\text{AIC}_x + \text{AIC}_y < \text{AIC}_0$, x_n and y_n are considered independent.

Chapter 8

1. Consider an AR model $y_n = \mu + a_1 y_{n-1} + \cdots + a_m y_{n-m} + v_n$ which has an additional parameter corresponding to the mean of the process.

2. Compare the following two models; the switched model $N(0, \sigma_1^1)$ for $n = 1, \ldots, k$ and $N(0, \sigma_2^1)$, and the stationary model $y_n \sim N(0, \sigma_0^2)$ for $n = 1, \ldots, N$. The AIC values of the models are respectively given by

$$
\begin{aligned}
\text{AIC}_{12} &= N \log 2\pi + k \log \hat{\sigma}^2 + (N-k) \log \hat{\sigma}_2^2 + N + 2 \times 2 \\
\text{AIC}_0 &= N \log 2\pi + N \log \hat{\sigma}_0^2 + N + 2.
\end{aligned}
$$

If $\text{AIC}_{12} < \text{AIC}_0$, it is considered that the variance changed at $n = k+1$.

3. Assume that $y_n = a_0 + a_1 x_n + \cdots + a_p x_n^p + \varepsilon_n$ for $n = 1, \ldots, k$, and that $y_n = b_0 + b_1 x_n + \cdots + a_q x_n^q + \delta_n$ for $n = k+1, \ldots, N$, where $\varepsilon_n \sim N(0, \sigma_1^2), \delta_n \sim N(0, \sigma_2^2)$.

$$
\text{AIC} = N \log 2\pi + k \log \hat{\sigma}_1^2 + (N-k) \log \hat{\sigma}_2^2 + N + 2 \times (p+q+4).
$$

4. To ensure continuity of the trend, require the parameters to satisfy

$$
a_0 + a_1 k + \cdots + a_p k^p = b_0 + b_1 k + \cdots + b_q k^q.
$$

To ensure smoothness of the trend, apply the further restriction

$$
a_0 + 2a_1 k + \cdots + p a_p k^p = b_0 + 2b_1 k + \cdots + q b_q k^q.
$$

Note that in computing AIC, the number of free parameters is decreased by 1 and 2, respectively.

5. The amount of computation in fitting an ordinary AR model is roughly $Nm^2/2$. On the other hand, the two locally stationary AR models need

$$\frac{1}{2}n_0 m^2 + \frac{n_1 - n_0}{p} \frac{1}{2}(p+1)m^2$$

$$\frac{1}{2}(N - n_1)m^2 + \frac{n_1 - n_0}{p} \frac{1}{2}(p+1)m^2,$$

respectively. Therefore, the total amount of computation required is evaluated as

$$\frac{1}{2}Nm^2 + \frac{1}{2}(n_1 - n_0)m^2 + \frac{n_1 - n_0}{p}m^2.$$

Even for $p = 1$, when the maximum of the sum of the second and third terms is attained with a value of $3/2(n_1 - n_0)m^2$, the increase in the amount of computation is at most three times that of the original AR model.

Chapter 9

1. Multiplying by T from the left on both sides of (9.1), we obtain $Tx_n = TF_n x_n + TG_n v_n = TF_n T^{-1} Tx_n + TG_n v_n$. Then, putting $z_n = Tx_n$ and using $y_n = H_n x_n + w_n = H_n T^{-1} Tx_n + w_n$, we can verify that we obtain an equivalent representation.

2.(1)

$$\text{Initial value} \quad : \quad x_{0|0} = 0, V_{0|0} = 100$$
$$\text{Prediction} \quad : \quad x_{n|n-1} = x_{n-1|n-1}, V_{n|n-1} = V_{n-1|n-1} + \tau^2$$
$$\text{Filter} \quad : \quad G_n = V_{n|n-1}/(V_{n|n-1} + 1),$$
$$x_{n|n} = x_{n|n-1} + G_n(y_n - x_{n|n-1}),$$
$$V_{n|n} = (1 - G_n)V_{n|n-1} = G_n.$$

(2) $V_{n+1|n} = V_{n|n-1}/(V_{n|n-1} + 1) + \tau^2$.

(3) In (2), put $V_{n+1|n} = V_{n|n-1} = V$.

(4) Denoting the solution of (3) by V, the variance of the filter is obtained by $V' = V/(V+1)$, and

$$x_{n+1|n} = x_{n|n} = x_{n|n-1} + \frac{V}{V+1}(y_n - x_{n|n-1}).$$

3. From $V_P = V$, $V_F = G = V/(V+1)$, $A = V/(V+1)/V = 1/(V+1)$ and $V_S = V_F + A^2(V_S - V_P)$, we obtain $V_S = V/(V+2)$.

τ^2	1	0.1	0.01	0.001
V_P	1.1618	0.370	0.105	0.032
V_F	0.618	0.270	0.095	0.031
V_S	0.447	0.156	0.050	0.016

Chapter 10

1. From $x_n = (y_n, y_{n-1}, \ldots, y_{n-m+1})^T$, the (i,j)-element of $E x_n x_n^T$ is $E y_{n-i+1} y_{n-j+1} = C_{i-j}$. Therefore, compute the autocovariance functions C_k, $k = 0, 1, \ldots$ of the AR model of order m and construct the initial matrix by

$$V_{0|0} = \begin{bmatrix} C_0 & C_1 & \cdots & C_{m-1} \\ C_1 & C_0 & \cdots & C_{m-2} \\ \vdots & \vdots & \ddots & \vdots \\ C_{m-1} & C_{m-2} & \cdots & C_0 \end{bmatrix}.$$

2. The change in the log-likelihood for the parameter transformed by (10.9) becomes very small if the absolute value of the parameter is large, and as a result, the optimization procedure may not work at all. Therefore, it is suggested that the range of the transformed parameter be restricted to some appropriately defined interval, say $[-20, 20]$.

Chapter 11

1.(1) For $y_{n+k} = y_{n+k-1} + v_{n+k} = \cdots = y_n + v_{n+1} + \cdots + v_{n+k}$, since $E v_{n+i} = 0$ and $E v_{n+i}^2 = \sigma^2$, we have $y_{n+k|n} = y_n$ and $E \varepsilon_{n+k|n}^2 = k\sigma^2$.

(2) From $k\sigma^2 = 4 \times 40,000 = 160,000$, we have $\sqrt{k}\sigma = 400$. For the normal distribution, the probability of a value larger than 1σ is 0.16.

(3) The variance σ^2 of the random walk changes over time. Therefore, we may consider a model taking into account changes in volatility

$$y_n = y_{n-1} + \sigma_n w_n,$$
$$\log \sigma_n^2 = a + b \log \sigma_{n-1}^2 + v_n \quad (or \ \log \sigma_n^2 = \log \sigma_{n-1}^2 + v_n).$$

2. We may, for example, consider a trigonometric model.
3. Omitted.

Chapter 12

1. If a characteristic root of the AR model approaches 1, discrimination between the AR model and the trend model becomes difficult. Therefore, it is recommended to apply some restriction so that the absolute value of the characteristic roots of the estimated AR model does not approach 1.

2. A seasonal component model based on a trigonometric function is obtained by $S_n = \sum_{j=1}^{6} (a_j \cos \frac{\pi j n}{6} + b_j \sin \frac{\pi j n}{6})$.

3.(1) Assuming that $\beta_{n1} = \beta_{n7}, \beta_{n2} = \cdots = \beta_{n6}$, we obtain a trading day effect with two parameters. In this case, the trading day effect is expressed by $td_n = \beta_{n1}(d'_{n1} + d'_{n7}) + \beta_{n2}(d'_{n2} + \cdots + d'_{n6})$.

 (2) Assuming that $\beta_{n2} = \cdots = \beta_{n6}$, we obtain a trading day effect model with three parameters.

4. The holiday pattern of Easter or "Golden week in Japan" changes year by year. Therefore, it is necessary to take account of these effects carefully. The effect of a leap year may not be completely removed by the standard trading day adjustment.

5. Omitted.

Chapter 13

1. (Example.) Stock-price data

2. In a time-varying coefficient regression model $y_n = \beta_{0n} + \beta_{1n}x_{n1} + \cdots + \beta_{mn}x_{nm} + \varepsilon_n$, consider, for example, a random walk model $\beta_{jn} = \beta_{j,n-1} + v_n$ for the coefficients $\beta_{0n}, \beta_{1n}, \ldots, \beta_{mn}$. This model can be expressed as a state-space model, so as in the case of time varying coefficient AR model, the parameter can be estimated by the Kalman filter.

Chapter 14

1. Compared with the Gaussian distribution, they have characteristics that the density is concentrated in the center and at the same time, they also have heavy tails. Therefore, they are adequate for handling sudden structural changes or outliers that occur with small probability.

2. For $y \sim N(0,1)$, put $z = y^2$, $x = \log z$. Then the probability that z takes a smaller value than t is given by

$$F(t) = \mathrm{Prob}(y^2 < t) = \frac{2}{\sqrt{2\pi}} \int_0^{\sqrt{t}} e^{-\frac{y^2}{2}} dy.$$

Differentiating with respect to t yields

$$f(t) = \frac{dF}{dt} = \frac{1}{\sqrt{2\pi}} t^{-\frac{1}{2}} e^{-\frac{t}{2}}$$

(a χ^2 distribution with 1 degree of freedom.) From (4.51), the distribution of the log-transform $x = h(z)$ is given by

$$g(x) = \left| \frac{dh^{-1}}{dx} \right| f(e^x) = e^x \frac{1}{\sqrt{2\pi}} (e^x)^{-\frac{1}{2}} e^{-\frac{e^x}{2}} = \frac{1}{\sqrt{2\pi}} e^{\frac{x}{2} - \frac{e^x}{2}}.$$

3. In the trend model, define the distribution of the system noise by

$$p_v(x) = \begin{cases} \delta(0) & \text{with probability } 1 - \alpha \\ N(0, \tau^2) & \text{with probability } \alpha, \end{cases}$$

where $\delta(0)$ is the δ function with mass at $x = 0$.

Chapter 15

1. Assuming that y_n is a time series and t_n is a trend, consider the model

$$\begin{aligned} t_n &= t_{n-1} + v_n, \\ \log \sigma_n^2 &= \log \sigma_{n-1}^2 + u_n, \\ y_n &= t_n + \sigma_n w_n, \end{aligned}$$

where $u_n \sim N(0, \tau_1^2)$, $v_n \sim N(0, \tau_2^2)$, $w_n \sim N(0, 1)$. Note that in this case, the observation model is nonlinear.

2. In Monte Carlo simulation, in general, we need 100 times as many particles to increase the accuracy by one digit.

3. Without any re-sampling step, the weights of many particles approach zero, and the accuracy of approximation is rapidly reduced.

4. If the number of particles is m, the number of data values is N, the dimension of the state is k and the lag is L, the fixed-interval smoother and the fixed-lag smoother need to store $m \times k \times N$ and $m \times k \times L$ data elements, respectively. Note that $N \gg L$.

Chapter 16

1. Assume that $f(x)$ is the density of a uniform distribution, $f(x) = 1$ for $0 \le x \le 1$, and 0 otherwise. Then since

$$g(x) = \int_{-\infty}^{\infty} f(y)f(x-y)dy = \int_0^1 f(x-y)dy = \begin{cases} x & 0 \le x \le 1 \\ 2 - x & 1 \le x \le 2 \\ 0 & \text{otherwise} \end{cases},$$

the density function of W is a triangle.

2. Assume that $y_0 = 1$, $v_n \equiv 0$; then $y_n = a^n$. Therefore, if we take m such that $a^m < c$ (for example $c = 10^{-3}$), then the effect of the initial value can be ignored. For $a = 0.9, 0.5, 0.1$, m is given by 66, 10, 5, respectively.

3. Using the observed time series, construct the state vector x_n and then by using random numbers, we can generate $x_{n+1}, x_{n+2}, \ldots$. To simulate an AR model using the state given in (9.5), put $x_n = (y_n, y_{n-1}, \ldots, y_{n-m+1})^T$.

Bibliography

[1] Akaike, H. (1968), "On the use of a linear model for the identification of feedback systems", *Ann. Inst. Statist. Math.*, **20**, 425–439.

[2] Akaike, H. (1969), "Fitting autoregressive models for prediction", *Ann. Inst. Statist. Math.*, **21**, 243–247.

[3] Akaike, H. (1971), "Autoregressive model fitting for control", *Ann. Inst. of Statist. Math.*, **23**, 163–180.

[4] Akaike, H. (1973), "Information theory and an extension of the maximum likelihood principle", in *Second International Symposium in Information Theory*, B.N. Petrov and F. Caski, eds., Budapest, Akademiai Kiado, 267–281.

[5] Akaike, H. (1974), "A new look at the statistical model identification", *IEEE Trans. Auto. Control*, **AC-19**, 716–723.

[6] Akaike, H. (1977), "On entropy maximization principle", in *Applications of Statistics*, P.R. Krishnaiah, ed., North Holland, The Netherlands, 27-41.

[7] Akaike, H. (1980a), "Likelihood and the Bayes procedure", in *Bayesian Statistics*, J.M. Bernardo, M.H. De Groot, D.V. Lindley and A.F.M. Smith, eds., University Press, Valencia, Spain, 143-166.

[8] Akaike, H. (1980b), "Seasonal adjustment by a Bayesian modeling", *J. Time Series Anal.*, **1**, 1–13.

[9] Akaike, H. and Ishiguro, M. (1983), "Comparative study of X-11 and Bayesian procedure of seasonal adjustment," *Applied Time Series Analysis of Economic Data*, U.S. Bureau of the Census.

[10] Akaike, H. and Nakagawa, T. (1989), "*Statistical Analysis and Control of Dynamic Systems*", Kluwer, Dordrecht.

[11] Alspach, D.L. and Sorenson, H.W. (1972), "Nonlinear Bayesian estimation using Gaussian sum approximations", *IEEE Trans, Auto. Control*, **17**, 439–447.

[12] Anderson, B.D.O. and Moore, J.B. (1979), *Optimal Filtering*, Prentice Hall, New Jersey.

[13] Ansley, C.F. (1979), "An algorithm for the exact likelihood of a mixed autoregressive moving average process", *Biometrika*, **66**, 59–65.

[14] Bartlett, M.S. (1946) "On the theoretical specification and sampling properties of autocorrelated time series", Symposium on Auto-correlation in Time Series, 27–41.

[15] Bierman G.J. (1977), *Factorization Methods for Discrete Sequential Estimation*, Academic Press, New York.

[16] Blackman, R.B. and Tukey, J.W. (1959), *The Measurement of Power Spectra from the Viewpoint of Communications Engineering*, Dover, New York.

[17] Bloomfield, P. (1976), *Fourier Analysis of Time Series: An Introduction*, Wiley, New York.

[18] Bolt, B.A. (1987), *Earthquakes*, W.H. Freeman, San Francisco.

[19] Box, G.E.P. and Cox, D.R. (1964), "An analysis of transformations," *Journal of the Royal Statist. Soc.*, **IB-26**, 211–252.

[20] Box, G.E.P., Hillmer, S.C. and Tiao, G.C. (1978) "Analysis and modeling of seasonal time series", in *Seasonal Analysis of Time Series*, Zellner, A., ed., US Bureau of the Census, *Economic Research Report ER-1*, 309–334.

[21] Box, G.E.P. and Jenkins, G.M. (1970), *Time Series Analysis, Forecasting and Control*, Holden-day, San Francisco.

[22] Brillinger, D.R (1974), "Fourier analysis of stationary processes", *Proc. IEEE*, **62**, 1628–1643.

[23] Brockwell, P.J. and Davis, R.A., (1991), *Time Series: Theory and Methods*, Second Edition, Springer-Verlag, New York.

[24] Broyden, C. G. (1970), "The convergence of a class of double-rank minimization algorithms", *Journal of the Institute of Mathematics and Its Applications*, **6**, 76–90

[25] Burg, J.P. (1967), "Maximum entropy spectral analysis", in *Proc. 37th Meeting of the Society of Exploration Geophysicists*, Peprinted in Modern Spectrum Anglysis, Childers, D.G., ed., IEEE Press, New York (1978), 34–39.

[26] Carlin, B.P., Polson, N.G. and Stoffer, D.S. (1992), "A Monte Carlo approach to nonnormal and nonlinear state space modeling", *J. Amer. Statist. Assoc.*, **75**, 493–500.

[27] Carter, C.K and Kohn, R. (1993), "A comparison of Markov chain Monte Carlo sampling schemes for linear state space models", in *Proceedings American Statistical Association Business and Economic Statistics Section*, 131–136.

[28] Cleveland, W.S. and Devlin, S.J. (1980), "Calendar effects in monthly time series; Detection by spectrum analysis and graphical methods", *J. Amer. Statist. Assoc.*, **75**, 487–496.

[29] Cleveland, W.S., Devlin, S.J. and Terpenning, I. (1982), "The SABL seasonal adjustment and calendar adjustment procedures", *Time Series Analysis: Theory and Practice*, **1**, 539–564.

[30] Doucet, A., de Freitas, N. and Gordon, N. (2001), *Sequential Monte Carlo Methods in Practice*, Springer, New York.

[31] Durbin, J. and Koopman, S.J. (2001), Time Series Analysis by State Space Methods, Oxford University Press, New York.

[32] Fletcher, R. (1980) *Practical Methods of Optimization*, **1**: Unconstrained optimization, Wiley & Sons, Chichester.

[33] Frühwirth-Schnatter, S. (1994), "Data augmentation and dynamic linear models", *J. Time Series Anal.*, **15**, 2, 183–202.

[34] Golub, G. (1965), "Numerical methods for solving linear least-square problems", *Num. Math.*, **7**, 206–216.

[35] Good, L.J. and Gaskins, J.R. (1980), "Density estimation and bump hunting by the penalized likelihood method exemplified by scattering and meteorite data", *J. Amer. Statist. Assoc.*, **75**, 42–73.

[36] Gordon, N.J., Salmond, D.J. and Smith, A.F.M. (1993), "Novel approach to nonlinear/non-Gaussian Bayesian state estimation", *IEE Proceedings–F*, **140**, 107–113.

[37] Harrison, P.J. and Stevens, C.F. (1976), "Bayesian forecasting", (with discussion), *J. Royal Statist. Soc.*, **B 38**, 205–247.

[38] Harvey, A. (1989), *Forecasting, structural time series models and the Kalman filter*, Cambridge University Press, Victoria, Australia.

[39] Hillmer, S.C. (1982), "Forecasting time series with trading day variation", *J. Forecasting*, **1**, 385–395.

[40] Huber, P.J. (1967), "The behavior of maximum likelihood estimates under nonstandard conditions", In *Proceedings of the fifth Berkley Symosium on Statistics*, **1**, 221–233.

[41] Jazwinski, A.H. (1970), *Stochastic Processes and Filtering Theory*, Academic Press, New York.

[42] Jenkins, G.M. and Watts, J.G. (1968), *Spectral Analysis and Its Applications*, Holden Day, San Franclsco.

[43] Jiang X.Q. and Kitagawa, G. (1993), "A time varying vector autoregressive modeling of nonstationary time series, *Signal Processing*, **33**, 315–331.

[44] Jones, R.H. (1980), "Maximum likelihood fitting of ARIMA models to time series with missing observations", *Technometrics*, **22**, 389–395.

[45] Kalman, R.E. (1960) A new approach to linear filtering and prediction problems", *Trans. Amer. Soc. Mech. Eng., J. Basic Engineering*, **82**, 35–45.

[46] Kaminski, P.G., Bryson, A.E. and Schmidt, S.F. (1971), "Discrete square root filtering: A survey of current technique", *lEEE Trans. Auto. Control*, **AC-16**, 727–735.

[47] Kitagawa, G. (1983), "Changing spectrum estimation", *J. Sound and Vibration*, **89(4)**, 443–445.

[48] Kitagawa, G. (1987), "Non-Gaussian state space modeling of nonstationary time series, (with discussion)", *J. Amer. Statist. Assoc.*, **82**, 1032–1063.

[49] Kitagawa, G. (1989), "Non-Gaussian seasonal adjustment", *Computers & Mathematics with Applications*, **18**, 503–514.

[50] Kitagawa, G. (1991), "A nonlinear smoothing method for time series analysis", *Statistica Sinica*, **1**, 371–388.

[51] Kitagawa, G. (1993), "A Monte-Carlo filtering and smoothing method for non-Gaussian nonlinear state space models", *Proceedings of the 2nd U.S.-Japan Joint Seminar on Statistical Time Series Analysis*, 110-131.

[52] Kitagawa, G. (1996), "Monte Carlo filter and smoother for non-Gaussian nonlinear state space models", *J. of Comp. and Graph. Statist.*, **5**, 1–25.

[53] Kitagawa, G. and Akaike, H. (1978), "A procedure for the modeling of nonstationary time series", *Ann. Inst. Statist. Math.*, **30-B**, 215–363.

[54] Kitagawa, G. and Gersch, W. (1984), "A smoothness priors-state space modeling of time series with trend and seasonality", *J. Amer. Statist. Assoc.*, **79**, 378–389.

[55] Kitagawa, G. and Gersch, W. (1985), "A smoothness priors time

varying AR coefficient modeling of nonstationary time series", *IEEE Trans. on Auto. Control*, **AC-30**, 48–56.

[56] Kitagawa, G. and Gersch, W. (1996), *Smoothness Priors Analysis of Time Series, Lecture Notes in Statistics*, **116**, Springer, New York.

[57] Knuth, D.E. (1997), *The Art of Computer Programming*, Third Edition, Addison-Wesley, New Jersey.

[58] Konishi, S. and Kitagawa, G. (2008), *Information Criteria and Statistical Modeling*, Springer, New York.

[59] Kozin, F. and F. Nakajima (1980), "The order determination problem for linear time-varying AR models", *IEEE Transactions on Automatic Control*, **AC-25**, 250–257.

[60] Kullback, S. and Leibler, R.A. (1951), "On information and sufficiency", *Ann. Math. Stat.*, **22**, 79–86.

[61] Levinson, N. (1947). "The Wiener RMS error criterion in filter design and prediction." *J. Math. Phys.*, **25**, 261–278

[62] Lewis, T.G. and Payene, W.H. (1973), "Generalized feedback shift register psudorandom number algorithm", *Journal of ACM*, **20**, 456–468.

[63] Matsumoto, M. and Kurita, Y. (1994), "Twisted GFSR Generators II", *ACM Transactions on Modeling and Computer Simulation*, **4(3)**, 254–266.

[64] Matsumoto, M. and Nishimura, T. (1998), "Mersenne twister: a 623-dimensionally equidistributed uniform pseudorandom number generator", *ACM Transactions on Modeling and Computer Simulation*, **8**, 3–30.

[65] Ohtsu, K., Peng, H. and Kitagawa, G. (2015), *Time Series Modeling for Analysis and Control: Advanced Autopilot and Monitoring Systems*, Springer, Tokyo.

[66] Ozaki, T. and Tong, H. (1975), "On the fitting of nonstationary autoregressive models in time series analysis", in *Proceedings 8th Hawaii Intnl. Conf. on System Sciences*, 224–246.

[67] Priestley, M.B. and Subba Rao, T. (1969), "A test for nonstationarity of time-series", *Journal of the Royal Statist. Soc.*, **B-31**, 140–149.

[68] Quenouille, M.H. (1949), "Approximate tests of correlation in time series", *Journal of the Royal Statist. Soc.*, **B-11**, 68–84.

[69] Sage, A.P. and Melsa, J.L. (1971), *Estimation Theory with Applications to Communication and Control*, McGraw-Hill, New York.

[70] Sakamoto, T., Ishiguro M. and Kitagawa, G. (1986), *Akaike Information Criterion Statistics*, D. Reidel, Dordreeht.

[71] Schnatter, S. (1992), "Integration-based Kalman-filtering for a dynamic generalized linear trend model", *Computational Statistics and Data Analysis*, **13**, 447–459.

[72] Shibata, R. (1976) "Selection of the order of an autoregressive model by Akaike's information criterion", *Biometrika*, **63**, 117–126.

[73] Shiskin, J., Young, A.H. and Musgrave, J.C.(1976), "The X-11 variant of the Census method II seasonal adjustment program", *Technical Paper*, **15**, Bureau of the Census, U.S. Dept of Commerce.

[74] Shumway, R.H. and Stoffer, D.S. (2000), *Time Series Analysis and Its Applications*, Springer, New York.

[75] Subba Rao, T. (1970), "The fitting of non-stationary time-series models with time-dependent parameters", *Journal of the Royal Statist. Soc.*, **B-32**, 312–322.

[76] Takanami, T. (1991), "ISM data 43-3-01: Seismograms of forechocks of 1982 Urakawa-Oki earthquake", *Ann. Inst. Statist. Math.*, **43**, 605.

[77] Takanami, T. and Kitagawa, G. (1991), "Estimation of the arrival times of seismic waves by multivariate time series model", *Ann. Inst. Statist. Math.*, **43**, 407–433.

[78] Tanizaki, H. (1993), *Nonlinear Filters: Estimation and Applications*, Springer-Verlag, New York.

[79] Wahba, G. (1980), "Automatic smoothing of the log periodogram", *J. Amer. Statist. Assoc.*, **75**, 122–132.

[80] West, M. (1981), "Robust sequential approximate Bayesian estimation", *J. Royal Statist. Soc.*, **B, 43**, 157–166.

[81] West, M. and Harrison, P.J. (1989), *Bayesian Forecasting and Dynamic Models, Springer Series in Statistics*, Springer-Verlag, Berlin.

[82] Whittaker, E.T. and Robinson, G. (1924), *Calculus of Observations, A Treasure on Numerical Calculations*, Blackie and Son, Lmtd., London, 303–306.

[83] Whittle, P. (1963) "On the fitting of multivariable autoregressions and the approximate canonical factorization of a spectral density matrix", *Biometrika*, **50**, 129–134.

[84] Whittle, P. (1965), "Recursive relations for predictors of non-stationary processes", *J. Royal Statist. Soc.*, **B 27**, 523–532.

Index

abrupt changes of coefficients, 225
addition of data, 87
AIC, 69, 70, 73, 139, 182
Akaike information criterion, 69,
    70, 73
amplitude spectrum, 105
AR coefficient, 91
AR model, 92, 113, 154, 274
arbitrary distribution, 272
ARMA model, 91, 155, 171
ARMA order, 91
atmospheric pressure, 6
autocorrelation function, 23, 24
autocovariance, 21
autocovariance function, 22, 23,
    94, 175
automatic partition, 139
autoregressive coefficient, 91, 113
autoregressive model, 92, 113
autoregressive moving average
    model, 91, 171
autoregressive order, 91
averaged periodogram, 46

backward AR model, 285
barometric pressure, 28
Bayes factor, 253
BFGS formula, 284
binomial distribution, 245
bivariate time series, 6
BLSALLFOOD data, 5, 201, 205
Box-Cox transformation, 11, 74
Box-Muller transformation, 269

Burg's algorithm, 120

Cauchy distribution, 56, 67, 238,
    241, 245, 262, 272
central limit theorem, 71
change point, 144
characteristic equation, 100
characteristic roots, 100
χ^2 distribution, 42, 56, 57, 272
Cholesky decomposition, 276
coherency, 106
conditional densities, 231
conditional distribution, 156
continuous time series, 7
covariance function, 19
covariance stationary, 21
cross-correlation function, 29
cross-covariance function, 29, 104
cross-spectral density function,
    105

density function, 55
description of time series, 9
DFP formula, 284
differencing, 11
discrete time series, 7
distribution function, 55, 273
divided model, 140
double exponential distribution,
    57, 214, 240, 241
double exponential random
    number, 272, 280